I0073275

TRANSACTIONS

AMERICAN PHILOSOPHICAL SOCIETY

HELD AT PHILADELPHIA

FOR PROMOTING USEFUL KNOWLEDGE

VOLUME XXI—NEW SERIES

PART II

ARTICLE II—*On the Construction of Isobaric Charts for High Levels in the Earth's Atmosphere and their Dynamic Significance.   By J. W. Sandström*

Philadelphia:
THE AMERICAN PHILOSOPHICAL SOCIETY
104 SOUTH FIFTH STREET
1906

ARTICLE II.

ON THE CONSTRUCTION OF ISOBARIC CHARTS FOR HIGH LEVELS IN THE EARTH'S ATMOSPHERE AND THEIR DYNAMIC SIGNIFICANCE.

(Plate VIII.)

BY J. W. SANDSTRÖM, STOCKHOLM, SWEDEN.

(Read April 14, 1905.)

I. INTRODUCTION.

The construction of isobaric charts for high levels has been attempted by several investigators in dynamic meteorology. I will here only mention:

(a) Teisserenc de Bort's attempt to draw such charts over the whole earth based on the isobars and isotherms at sealevel, the observed direction of motion of the clouds, and an assumed probable diminution of temperature with altitude;

(b) Koeppen's graphic presentation of such charts based on the isobars and isoterms at sealevel, and

(c) Hergesell's construction of similar charts on the basis of the results of the international balloon ascensions.

From the relation of the isobaric charts for sealevel to the dynamics of the lower atmospheric strata, the analogous relation of the isobaric charts for higher levels to the dynamics of the upper strata has been correctly appreciated. Indeed from the charts already drawn we have succeeded in explaining many of the phenomena of the upper layers of the atmosphere, for example, the general circulation from West to East [*] and the movements of the clouds in the upper portions of cyclones.[†]

My attempts to apply Bjerknes' theory of solenoids [‡] to dynamic meteorology have led me also to the construction of isobaric charts for higher levels. This theory requires, however, that such charts be drawn on level surfaces of gravity and not on surfaces of equal elevation above sealevel. In the following pages I shall show how such charts can be constructed from meteorological observations obtained by means of kites and balloons in the free air.

---

[*] L. Teisserenc de Bort: Étude sur la circulation generale de l'atmosphere. Annales du Bureau Central Meteorologique de France, 1885, Tome 4.

[†] W. Koeppen: Ueber die Gestalt der Isobaren in ihrer Abhängung von Seehöhe u. Temperaturvertheilung. Met. Zeit., 1888, p. 476.

[‡] See Bjerknes, in Monthly Weather Review, 1900, October, pp. 434–443, December, pp. 532–535. Sandström: On the Application of Prof. V. Bjerknes' Theory, in Memoirs Royal Swedish Academy, 1900, vol. 33.

I shall then draw auxiliary charts that show the differences of pressure for any vertical line between sealevel and the higher levels; by a simple graphic superposition of these charts upon the isobaric charts drawn in the ordinary way for sealevel we shall obtain the isobaric charts for the various higher levels. It is necessary to proceed in this way in the construction because the kite and balloon stations are too far apart from each other to allow us to draw the upper isobars directly from the results obtained from the ascensions. On the other hand these kite and balloon results suffice quite well for drawing the charts of differences, because the differences change but little from place to place.

Furthermore, Bjerknes' theory leads to the construction of yet another kind of charts, namely those which represent the lines of intersection of any given isobaric surface with the level surfaces of gravity, and which are thus a kind of topographic charts of the different isobaric surfaces. These charts, which are closely related to the isobaric maps, are like those constructed by the superposition of difference-charts based on the observations made at fixed meteorological stations combined with those made by means of kites and balloons.

If the isobaric chart for any level not too far removed from sealevel is compared with the chart of isobars at sealevel, both charts will be found to show nearly the same type of isobars, and one can scarcely learn more from both together than from the chart for sealevel alone. In such a case, however, the difference-chart furnishes a much more effective means of discovering the relation between these two isobaric charts. Now it has been found that such difference-charts are very closely related to the Bjerknes' solenoids, so that indeed, the number and positions of the solenoids in the atmosphere are fully presented by these difference-charts. I shall therefore in this essay consider equally the difference-charts, the isobaric maps, and the topographic charts of isobaric surfaces.

I shall first construct the level surfaces of gravity in the atmosphere and then calculate the mutual positions of the isobaric surfaces and the level surfaces of gravity under both static and dynamic conditions. Thus all the aids necessary for the construction of the above-mentioned maps will be obtained. Finally I shall show how Bjerknes' theory is to be applied to these charts.

I would express my warmest thanks to the United States Weather Bureau for the abundant observational data so kindly sent me. I also owe many thanks to Professor V. Bjerknes for his interest and many good suggestions and the support which he has given me during the progress of my work.

II. The Level Surfaces of Gravity.

We first consider the level surfaces of gravity because, by reason of their absolutely fixed positions with relation to the earth, they are specially adapted to serve as coördinate planes in the atmosphere. Let it be remarked in passing, that all the burdensome corrections in meteorological work arising from the variations of gravity with elevation and geographical latitude disappear * if once for all we introduce level surfaces of gravity as the coördinate planes in place of surfaces of equal elevation above sealevel.

The level surfaces of gravity are surfaces which are at every point perpendicular to the direction of the gravitational force.† A fundamental property of the level surfaces of gravity results directly from this definition, viz.: no work is necessary to shift a mass from any point in a level surface to any other point in the same surface. Further it also follows that the same amount of work must be performed to transfer a mass from any given level surface to any other given level surface, quite independently of the path along which the transfer takes place. We shall make use of this property in the construction of our system of level surfaces in the atmosphere by choosing the surface of sealevel [i. e., the geodesist's spheroid], as our zero-surface and distributing the other surfaces in such a way that it will always require just one unit of work to raise the unit of mass from one level surface to the surface next above it. As unit of mass we choose 1 pound (English) and as unit of work one $\dfrac{\text{pound} \times \text{mile}^2}{\text{hour}^2}$.

To raise one pound through the vertical distance of one mile requires a number g of units of work, if by g we indicate the acceleration of gravity in mile/hour^2. If

---

* This does not refer to the reduction of the mercurial barometer to normal gravity, because this is to be considered as an instrumental correction.

† Note by the Editor : This is the so-called "apparent gravity" or the attraction of the earth as diminished by the distance from the earth's center and also by the centrifugal force due to the diurnal rotation of the globe.

Let the term geoid apply to the natural irregular surface of the earth and the term spheroid to the ideal regular surface of the geodesist which coincides nearly with sealevel and is necessarily a level surface. The observed values of acceleration of apparent gravity made at points on the surface of the geoid are usually reduced vertically downward to a point on the ideal spheroid by some one of several formulæ, and the collation of all such reduced values shows that for this spheroid in general

$$g = 32.172\ 6\ (1 - 0.002\ 59 \cos 2\lambda).$$

For a point on the geoid surface, h in feet, or H in meters, above this spheroid apparent gravity diminishes by distance but increases by the attraction of the intervening earth, as represented altogether by the factor $\left(1 - \dfrac{5}{4} \cdot \dfrac{H}{R}\right)$, i. e.,

$$(1 - 0.000\ 000\ 059\ 7h) \quad \text{or} \quad (1 - 0.000\ 000\ 196H)$$

For a point in the atmosphere, z in feet or Z in meters, above the geoid surface apparent gravity diminishes by increase of distance only, or by the factor $(1 - 2z/R)$, i. e., $(1 - 0.000\ 000\ 095\ 7z)$ or $(1 - 0.000\ 000\ 314Z)$. Hence starting from the geoid surface we may say that apparent gravity increases with descent by the factor $\left(1 + \dfrac{5}{4} \cdot \dfrac{H}{R}\right)$, but decreases with ascent by the factor $(1 - 2z/R)$.

however, g be expressed in feet/second^2 units as is customarily done, then we find that in order to raise one pound a vertical distance of one foot the expression

$$0.464\ 876 \times g \times \frac{\text{pound} \times \text{mile}^2}{\text{hour}^2}$$

represents the amount of work which must be performed. Therefore every foot of vertical distance will be intersected by $0.464\ 876 \times g$ level surfaces of gravity. At the Equator, where gravity equals $32.089\ \frac{\text{feet}}{\text{sec.}^2}$, there will be $0.464\ 876 \times 32.089 = 14.917$ such planes; and at either pole, where gravity equals 32.256, there will be $0.464\ 876 \times 32.256 = 14.995$ such planes to every foot of vertical rise. These figures hold true near sealevel, while at greater heights the level surfaces will lie somewhat farther apart. The level surfaces are thus seen to constitute closed surfaces at approximately one-fifteenth foot intervals from one another, enclosing the earth and showing a polar flattening similar to that of the ocean surface.

In order to distinguish the individual surfaces of this system they are numbered as follows: sealevel is numbered zero (0); the plane standing about one-fifteenth foot above zero is numbered one (1); the plane standing about two-fifteenths foot above zero is numbered two (2) and so onward. Thus the surface numbered ten (10) has an elevation of about two-thirds foot; number 100 an elevation of about 6⅔ feet; the planes numbered 1 000, 10 000, 100 000, etc., have respectively heights of about 67, 669, 6 690 feet, etc., above sealevel. The true elevations above sealevel of these level surfaces are somewhat greater at the Equator and somewhat less at the poles, than the average values here given.

If now these level surfaces of gravity are to be used as coördinate surfaces in the atmosphere instead of the surfaces of equal elevation above sealevel, then instead of expressing the elevation of any point in feet above sealevel we must state the ordinal number of the level surface in which it lies. The transformation from "feet above sealevel" to the ordinal number of the level surface of gravity may be easily performed by means of a table showing the relation between the two numbers. Such a table should be calculated for every locality where the elevations of kites, balloons or clouds are measured, and in the following paragraphs I show how such a table may be calculated.

Designate the elevation above sealevel of the point by z, and the ordinal number of the level surface in which it lies by V. Then V is equal to the number of level surfaces included between the given point and sealevel. V also expresses the work required to be done in order to raise a unit mass from sealevel to the position of the given point, for it always requires one unit of work to raise a unit mass from one surface to the next higher one. Now this total quantity of work required is equal to

$$\int_0^z g \cdot dz$$

where by dz we designate an element of the vertical line from the point to sealevel and by g designate the accleration of gravity for this element. We thus obtain the following relation between V, g, and z:

$$V = \int_0^z g \cdot dz \tag{1}$$

where the integration is to be carried out along the vertical line joining the point with sealevel. The distribution of gravity along the vertical and above the surface of the earth is given by the well known formula

$$g = g_0(1 - 0.000\ 000\ 095\ 7(z - z_0)), \tag{2}$$

where z_0 represents the elevation of the earth's surface above sealevel, and g_0 is the acceleration of gravity at the earth's surface. If z represents depth below the earth's surface then g at this depth is given by the formula

$$g = g_0(1 + 0.000\ 000\ 059\ 7(z_0 - z)). \tag{3}$$

Here and in what follows, by the earth's surface in the neighborhood of a meteorological station is always meant the level of the barometer of the station, or the level from which cloud-altitudes, kite-altitudes and the like are calculated [i. e., the so-called "station level" of the United States Weather Bureau].

The ordinal number V_0 of the gravity surface which coincides with the surface of the earth at the station is obtained by substituting equation (3) in equation (1) and integrating from sealevel up to the surface of the earth. We thus find

$$V_0 = 0.464\ 876 \times g_0 \times z_0(1 + 0.000\ 000\ 029\ 85 z_0). \tag{4}$$

For example, to find V_0 for the kite-station at Omaha, Nebr., we substitute the altitude above sealevel, $z_0 = 1\ 241$ feet, and the acceleration of gravity at the earth's surface at Omaha, $g_0 = 32.160$ foot/sec.², in formula (4); whence we have

$$V_0 = 0.464\ 876 \times 32.160 \times 1\ 241(1 + 0.\ 000\ 000\ 029\ 85 \times 1\ 241)$$
$$= 18\ 550.$$

There are thus seen to be 18 550 level surfaces of gravity between sealevel and the level of the barometer of the kite-station at Omaha; or work to the amount of

$$18\ 550 \frac{\text{pound} \times \text{mile}^2}{\text{hour}^2}$$

must be performed in order to raise one pound from sealevel to the level of the station barometer in Omaha. From now on the numbers of these level surfaces of gravity will be expressed in even tens, since the heights are not measured closer than to one foot.

If now we substitute in (1) the value of gravity obtained from (2) and continue the integration from the surface of the earth up to the elevation z above sealevel, we obtain the ordinal number V of the level surface that passes through the point at the elevation z. We find

$$V = V_0 + 0.464\ 876 g_0(z - z_0)(1 - 0.000\ 000\ 047\ 85(z - z_0))$$

where $z - z_0$ is the elevation of the point above the earth's surface. If this elevation, $z - z_0$, be represented by z_1 then we have

$$V = V_0 + 0.464\ 876 g_0 z_1(1 - 0.000\ 000\ 047\ 85 z_1). \tag{5}$$

The calculation of V is much simplified by using the small Table I, which contains the value of the quantity $0.464\ 876 \times z_1(1 - 0.000\ 000\ 047\ 85 z_1)$ for each 1 000 feet of elevation above the earth's surface.

TABLE 1.

$0.464\ 876 \times z_1(1 - 0.000\ 000\ 047\ 85z_1).$

z_1	$(V - V_0)/g_0$
1 000 ft.	464.85
2 000	929.66
3 000	1 394.43
4 000	1 859.15
5 000	2 323.82
6 000	2 788.46
7 000	3 253.04
8 000	3 717.58
9 000	4 182.08
10 000	4 646.54

Thus to calculate V for Omaha, we must, according to formula (5) multiply the values given in Table 1 by $g_0 = 32.160$ and then add the quantities thus obtained to $V_0 = 18\ 550$. We thus obtain the values given in Table 2.

.TABLE 2.

GRAVITY-POTENTIAL TABLE FOR OMAHA, NEBR.

z_1	V
1 000 ft.	33 500
2 000	48 448
3 000	63 395
4 000	78 340
5 000	93 285
6 000	108 227
7 000	123 168
8 000	138 107
9 000	153 046
10 000	167 983

By linear interpolation in this Table we obtain Table 3 which we may designate as the gravity-potential table for Omaha, since V is identical with the potential of gravity according to (1). In other words by taking the derivative of that formula we find

$$\frac{dV}{dz} = g.$$

CONSTRUCTION OF ISOBARIC CHARTS

TABLE 3.

TABLE OF GRAVITY POTENTIALS FOR OMAHA, NEBR.

z_1	0	10	20	30	40	50	60	70	80	90
0	18550	18700	18850	19000	19150	19300	19450	19600	19750	19900
100	20050	20190	20340	20490	20640	20790	20940	21090	21240	21390
200	21540	21690	21840	21990	22140	22290	22440	22590	22740	22890
300	23040	23180	23330	23480	23630	23780	23930	24080	24230	24380
400	24530	24680	24830	24980	25130	25280	25430	25580	25730	25880
500	26030	26170	26320	26470	26620	26770	26920	27070	27220	27370
600	27520	27670	27820	27970	28120	28270	28420	28570	28720	28870
700	29020	29160	29310	29460	29610	29760	29910	30060	30210	30360
800	30510	30660	30810	30960	31110	31260	31410	31560	31710	31860
900	32010	32150	32300	32450	32600	32750	32900	33050	33200	33350
1000	33500	33650	33800	33950	34100	34250	34400	34550	34700	34850
1100	35000	35140	35290	35440	35590	35740	35890	36040	36190	36340
1200	36490	36640	36790	36940	37090	37240	37390	37540	37690	37840
1300	37990	38130	38280	38430	38580	38730	38880	39030	39180	39330
1400	39480	39630	39780	39930	40080	40230	40380	40530	40680	40830
1500	40980	41120	41270	41420	41570	41720	41870	42020	42170	42320
1600	42470	42620	42770	42920	43070	43220	43370	43520	43670	43820
1700	43970	44110	44260	44410	44560	44710	44860	45010	45160	45310
1800	45460	45610	45760	45910	46060	46210	46360	46510	46660	46810
1900	46960	47100	47250	47400	47550	47700	47850	48000	48150	48300
2000	48450	48600	48750	48900	49050	49200	49350	49500	49650	49800
2100	49940	50090	50240	50390	50540	50690	50840	50990	51140	51290
2200	51440	51590	51740	51890	52040	52190	52340	52490	52640	52790
2300	52930	53080	53230	53380	53530	53680	53830	53980	54130	54280
2400	54430	54580	54730	54880	55030	55180	55330	55480	55630	55780
2500	55920	56070	56220	56370	56520	56670	56820	56970	57120	57270
2600	57420	57570	57720	57870	58020	58170	58320	58470	58620	58770
2700	58910	59060	59210	59360	59510	59660	59810	59960	60110	60260
2800	60410	60560	60710	60860	61010	61160	61310	61460	61610	61760
2900	61900	62050	62200	62350	62500	62650	62800	62950	63100	63250
3000	63400	63550	63700	63850	64000	64150	64300	64450	64600	64740
3100	64890	65040	65190	65340	65490	65640	65790	65940	66090	66240
3200	66390	66540	66690	66840	66990	67140	67280	67430	67580	67730
3300	67880	68030	68180	68330	68480	68630	68780	68930	69080	69230
3400	69380	69530	69670	69820	69970	70120	70270	70420	70570	70720
3500	70870	71020	71170	71320	71470	71620	71770	71920	72070	72210
3600	72360	72510	72660	72810	72960	73110	73260	73410	73560	73710
3700	73860	74010	74160	74310	74460	74610	74750	74900	75050	75200
3800	75350	75500	75650	75800	75950	76100	76250	76400	76550	76700
3900	76850	77000	77140	77290	77440	77590	77740	77890	78040	78190
4000	78340	78490	78640	78790	78940	79090	79240	79390	79540	79690
4100	79840	79980	80130	80280	80430	80580	80730	80880	81030	81180
4200	81330	81480	81630	81780	81930	82080	82230	82380	82530	82680
4300	82830	82970	83120	83270	83420	83570	83720	83870	84020	84170
4400	84320	84470	84620	84770	84920	85070	85220	85370	85520	85670
4500	85820	85960	86110	86260	86410	86560	86710	86860	87010	87160
4600	87310	87460	87610	87760	87910	88060	88210	88360	88510	88660
4700	88810	88950	89100	89250	89400	89550	89700	89850	90000	90150
4800	90300	90450	90600	90750	90900	91050	91200	91350	91500	91650
4900	91800	91940	92090	92240	92390	92540	92690	92840	92990	93140

P.P.

1	10
2	30
3	40
4	60
5	70
6	90
7	100
8	120
9	130

TABLE 3 (*Concluded*).

TABLE OF GRAVITY POTENTIALS FOR OMAHA, NEBR.

z_1	0	10	20	30	40	50	60	70	80	90
5000	93290	93440	93590	93740	93890	94040	94190	94340	94490	94630
5100	94780	94930	95080	95230	95380	95530	95680	95830	95980	96130
5200	96280	96430	96580	96730	96880	97030	97170	97320	97470	97620
5300	97770	97920	98070	98220	98370	98520	98670	98820	98970	99120
5400	99270	99420	99560	99710	99860	100010	100160	100310	100460	100610
5500	100760	100910	101060	101210	101360	101510	101660	101810	101960	102100
5600	102250	102400	102550	102700	102850	103000	103150	103300	103450	103600
5700	103750	103900	104050	104200	104350	104500	104640	104790	104940	105090
5800	105240	105390	105540	105690	105840	105990	106140	106290	106440	106590
5900	106740	106890	107030	107180	107330	107480	107630	107780	107930	108080
6000	108230	108380	108530	108680	108830	108980	109130	109280	109430	109570
6100	109720	109870	110020	110170	110320	110470	110620	110770	110920	111070
6200	111220	111370	111520	111670	111820	111970	112110	112260	112410	112560
6300	112710	112860	113010	113160	113310	113460	113610	113760	113910	114060
6400	114210	114360	114500	114650	114800	114950	115100	115250	115400	115550
6500	115700	115850	116000	116150	116300	116450	116600	116750	116900	117040
6600	117190	117340	117490	117640	117790	117940	118090	118240	118390	118540
6700	118690	118840	118990	119140	119290	119440	119580	119730	119880	120030
6800	120180	120330	120480	120630	120780	120930	121080	121230	121380	121530
6900	121680	121830	121970	122120	122270	122420	122570	122720	122870	123020
7000	123170	123320	123470	123620	123770	123920	124070	124220	124370	124510
7100	124660	124810	124960	125110	125260	125410	125560	125710	125860	126010
7200	126160	126310	126460	126610	126760	126910	127050	127200	127350	127500
7300	127650	127800	127950	128100	128250	128400	128550	128700	128850	129000
7400	129150	129300	129440	129590	129740	129890	130040	130190	130340	130490
7500	130640	130790	130940	131090	131240	131390	131540	131690	131840	131980
7600	132130	132280	132430	132580	132730	132880	133030	133180	133330	133480
7700	133630	133780	133930	134080	134230	134380	134520	134670	134820	134970
7800	135120	135270	135420	135570	135720	135870	136020	136170	136320	136470
7900	136620	136770	136910	137060	137210	137360	137510	137660	137810	137960
8000	138110	138260	138410	138560	138710	138860	139010	139160	139310	139450
8100	139600	139750	139900	140050	140200	140350	140500	140650	140800	140950
8200	141100	141250	141400	141550	141700	141850	141990	142140	142290	142440
8300	142590	142740	142890	143040	143190	143340	143490	143640	143790	143940
8400	144090	144240	144380	144530	144680	144830	144980	145130	145280	145430
8500	145580	145730	145880	146030	146180	146330	146480	146630	146780	146920
8600	147070	147220	147370	147520	147670	147820	147970	148120	148270	148420
8700	148570	148720	148870	149020	149170	149320	149460	149610	149760	149910
8800	150060	150210	150360	150510	150660	150810	150960	151110	151260	151410
8900	151560	151710	151850	152000	152150	152300	152450	152600	152750	152900
9000	153050	153200	153350	153500	153650	153800	153950	154100	154240	154390
9100	154540	154690	154840	154990	155140	155290	155440	155590	155740	155890
9200	156040	156190	156330	156480	156630	156780	156930	157080	157230	157380
9300	157530	157680	157830	157980	158130	158280	158420	158570	158720	158870
9400	159020	159170	159320	159470	159620	159770	159920	160070	160220	160370
9500	160520	160660	160810	160960	161110	161260	161410	161560	161710	161860
9600	162010	162160	162310	162460	162610	162750	162900	163050	163200	163350
9700	163500	163650	163800	163950	164100	164250	164400	164550	164700	164840
9800	164990	165140	165290	165440	165590	165740	165890	166040	166190	166340
9900	166490	166640	166790	166930	167080	167230	167380	167530	167680	167830

P.P.

1	10
2	30
3	40
4	60
5	70
6	90
7	100
8	120
9	130

A. P. S.—XXI. B. 21, 11, '05.

The dimension or "dimensional equation" of the quantity V is obtained from formula (1), and in fact this quantity is expressed in terms of $\dfrac{\text{distance}^2}{\text{time}^2}$, that is to say the dimension for work done upon a unit mass. In Table 3 the unit for V is chosen as one $\dfrac{\text{mile}^2}{\text{hour}^2}$ in order that the velocities resulting from the solenoids may be expressed in $\dfrac{\text{mile}}{\text{hour}}$. In order to obtain from Table 3 the value of V at any given elevation, e. g., 3 487 feet, above the level of the station barometer at Omaha, we proceed as follows. First in the principal Table 3 we seek the value of V corresponding to $z = 3\,480$ feet, viz., 70 570; then by the aid of the small auxiliary table of proportional parts we find for $z = 7$ feet the additional portion of $V = 100$ and thus the complete $V = 70\,670$ for $z = 3\,487$ feet. Consequently work amounting to 70 670 $\dfrac{\text{mile}^2}{\text{hour}^2}$ must be performed in order to raise the unit mass from sealevel to the altitude of 3 487 feet above the station barometer at Omaha, or we may say that there are 70 670 level surfaces of gravity between sealevel and the point standing 3 487 feet above the Omaha station barometer.

This method for the calculation of V can be applied at all stations where g_0 has been previously determined by pendulum observations. At points where no such measurements of g_0 have been made the following well-known formula for the calculation of gravity at the earth's surface must be employed,

$$g_0 = 32.1726(1 - 0.002\,59 \cos 2\lambda)(1 - 0.000\,000\,059\,7z_0). \qquad (6)$$

TABLE 4.

THE ACCELERATION OF GRAVITY AT SEALEVEL.

Latitude.	0°	1°	2°	3°	4°	5°	6°	7°	8°	9°
0°	32.089	32.089	32.089	32.090	32.090	32.091	32.091	32.092	32.092	32.093
10	.094	.095	.096	.098	.099	.100	.102	.104	.105	.107
20	.109	.111	.113	.115	.117	.119	.121	.124	.126	.128
30	.131	.133	.136	.139	.141	.144	.147	.150	.152	.155
40	.158	.161	.164	.167	.170	.173	.176	.178	.181	.184
50	.187	.190	.193	.196	.198	.201	.204	.206	.209	.212
60	.214	.217	.219	.222	.224	.226	.228	.231	.233	.235
70	.236	.238	.240	.242	.243	.245	.246	.247	.249	.250
80	.251	.252	.253	.253	.254	.255	.255	.255	.256	.256

TABLE 5.

DECREASE OF GRAVITY WITH ELEVATION ABOVE SEALEVEL.

Elevation.	Decrease.
1 000 ft.	−0.002
2 000	−0.004
3 000	−0.006
4 000	−0.008
5 000	−0.010
6 000	−0.012
7 000	−0.013
8 000	−0.015
9 000	−0.017
10 000	−0.019

Table 4 shows the acceleration of gravity at sealevel, and Table 5 the decrease in the acceleration of gravity with elevation above sealevel calculated according to formula (6). To find the value of g at the surface of the earth, for instance at Omaha, by the aid of these tables one first seeks in Table 4 for the value of g at sealevel for the latitude of Omaha ($\lambda = 41° 16'$) and finds it to be 32.162. From this value one then subtracts the correction 0.002 given in Table 5 for the elevation ($z = 1\ 241$ feet) above sealevel; hence the value 32.160 for g at the surface of the earth [$i.\ e.$, the geoid] at Omaha.

When one would consider the influence of the topography of the earth's surface on the dynamic meteorological processes he constructs charts having lines of equal values of V_0 instead of contour lines of equal elevation above sealevel. Such charts of lines of V_0 may be easily constructed from the contour charts by means of Table 6, which gives the elevations above sealevel of the lines $V_0 = 10\ 000$, $V_0 = 20\ 000$, etc., to $V_0 = 150\ 000$, for each 10° of latitude, north or south.

TABLE 6.

ELEVATIONS ABOVE SEALEVEL OF THE V_0 SURFACES FOR EACH TEN DEGREES OF LATITUDE.

V_0	0°	10°	20°	30°	40°	50°	60°	70°	80°	90°
10000	670	670	670	669	669	668	668	667	667	667
20000	1341	1341	1340	1339	1338	1337	1336	1335	1334	1334
30000	2011	2011	2010	2009	2007	2005	2003	2002	2001	2001
40000	2682	2681	2680	2678	2676	2673	2671	2669	2668	2668
50000	3352	3352	3350	3348	3345	3342	3339	3337	3335	3335
60000	4023	4022	4020	4017	4014	4010	4007	4004	4002	4002
70000	4693	4692	4690	4687	4683	4679	4675	4672	4670	4669
80000	5364	5363	5360	5357	5352	5347	5343	5339	5337	5336
90000	6034	6033	6031	6026	6021	6016	6011	6007	6004	6003
100000	6705	6704	6701	6696	6690	6684	6679	6674	6671	6670
110000	7375	7374	7371	7366	7360	7353	7347	7342	7338	7337
120000	8046	8045	8041	8036	8029	8022	8015	8009	8006	8005
130000	8717	8715	8711	8706	8698	8690	8683	8677	8673	8672
140000	9388	9386	9382	9375	9367	9359	9351	9345	9341	9339
150000	10058	10057	10052	10045	10037	10028	10019	10012	10008	10006

Such a map for North America, constructed by the aid of this table, is shown in Pl. VIII. The curves of V_0 on this map show that by reason of gravitation it always requires the performance of work amounting to 10 000 $\frac{\text{mile}^2}{\text{hour}^2}$ in order to raise the unit mass from a point on one curve to any point on the curve next above.

III. The Relative Positions of the Isobaric Surfaces and the Level Surfaces of Gravity under Static Conditions.

The well-known condition for atmospheric equilibrium is that the isobaric surfaces and the level surfaces of gravity shall coincide. If this condition is fulfilled then we may express the pressure p as a function of the gravity-potential only; and conversely can write the gravity-potential V as a function of the pressure only. In the following pages pressure considered as a function of gravity-potential will be represented by p_V, and gravity-potential considered as a function of pressure will be represented by V_p. The values of these functions are obtained by integrating the differential equation for the barometric determination of heights.* Since it is convenient to perform these integrations at first for special intervals, the following expressions are introduced:

$$\mathrm{E}_{p_0}^{p_1} = V_{p_1} - V_{p_0} \tag{7}$$

$$\Pi_{V_0}^{V_1} = p_{V_0} - p_{V_1} . \tag{8}$$

According to the above given definitions the quantities V_{p_0} and V_{p_1} are equal to the number of level surfaces of gravity expressed in $\frac{\text{mile}^2}{\text{hour}^2}$ units lying between sealevel and the isobaric surfaces p_0 and p_1 respectively; and $\mathrm{E}_{p_0}^{p_1}$ is the number of level surfaces between the two isobaric surfaces p_0 and p_1. The quantities p_{V_0} and p_{V_1} are the numbers of isobaric surfaces lying between sealevel and the two level surfaces of gravity numbered V_0 and V_1 respectively. $\Pi_{V_0}^{V_1}$ represents the number of isobaric surfaces lying between the two level surfaces of gravity V_0 and V_1. In all this we imagine the existence in the atmosphere of an isobaric surface for each inch of the column of a mercurial barometer [under standard gravity].

To calculate $\mathrm{E}_{p_0}^{p_1}$ we start with the equation of condition for dry air, viz.:

$$\frac{pv}{T} = \frac{p_0 v_0}{T_0} \tag{9}$$

and with the differential equation for the barometric measurement of altitudes, viz.:

---

* Note by the Editor : All barometric readings and isobars refer to absolute pressures as indicated by the mercurial column reduced to standard temperature, gravity, etc.

$$gdz = -vdp. \tag{10}$$

By solving (9) for v and substituting in (10) we obtain

$$gdz = -\frac{p_0 v_0}{T_0} \times T \times \frac{dp}{p}. \tag{11}$$

But from (1) we see that

$$dV = gdz,$$

and if we substitute this in (11) we have

$$dV = -\frac{p_0 v_0}{T_0} \cdot T \cdot \frac{dp}{p}. \tag{12}$$

By integrating formula (12) from $p = p_0$ to $p = p_1$ we obtain

$$V_{p_1} - V_{p_0} = \frac{p_0 v_0}{T_0} \int_{p_1}^{p_0} T \cdot \frac{dp}{p} \tag{13}$$

or by substituting from equation (7)

$$\mathrm{E}_{p_0}^{p_1} = \frac{p_0 v_0}{T_0} \int_{p_1}^{p_0} T \cdot \frac{dp}{p}. \tag{14}$$

In the calculation of $\Pi_{V_0}^{V_1}$ we may start with equation (12). First solving for $\frac{dp}{p}$ and then integrating from $V = V_0$ to $V = V_1$ we obtain

$$\log \text{ nat. } \frac{p_{V_1}}{p_{V_0}} = -\frac{T_0}{p_0 v_0} \int_{V_0}^{V_1} \frac{dV}{T}$$

or

$$p_{V_0} - p_{V_1} = p_{V_0}(1 - e^{-\frac{T_0}{p_0 v_0} \int_{V_0}^{V_1} \frac{dV}{T}}) \tag{15}$$

whence by (8) we find

$$\Pi_{V_0}^{V_1} = p_{V_0}(1 - e^{-\frac{T_0}{p_0 v_0} \int_{V_0}^{V_1} \frac{dV}{T}}). \tag{16}$$

Now by substituting the values

$$p_0 = 2.4934 \times 32.1726 \times 846.728,$$

$$v_0 = 1/0.080\ 259,$$

$$T_0 = 459.4 + 32.0 = 491.4,$$

in equation (14) we obtain the following expression

$$\mathrm{E}_{p_0}^{p_1} = \frac{2.4934 \times 32.1726 \times 846.728}{0.080259 \times 491.4} \int_{p_1}^{p_0} T \frac{dp}{p}. \tag{17}$$

The dimension of this expression is most readily found when it is written in the following form

$$E_{p_0}^{p_1} = 2.4934 \times 32.1726 \times \frac{846.728}{0.080\ 529} \int_{p_1}^{p_0} \frac{T}{491.4} \cdot \frac{dp}{p}.$$

In this expression the quantity 2.4934 is the height in feet of the mercurial column for a pressure of one atmosphere, and hence it has the dimension, foot. The number 32.1726 is the acceleration of gravity at sealevel at latitude 45° and has the dimension $\frac{\text{foot}}{\text{second}^2}$. The quotient $\frac{846.728}{0.080\ 529}$ is the ratio of the densities of mercury and air and has the dimension zero. The two remaining quotients, $\frac{T}{491.4}$ and $\frac{dp}{p}$ are also non-dimensional. Therefore the dimension of the whole expression is $\frac{\text{foot}^2}{\text{second}^2}$. In order to convert this into $\frac{\text{mile}^2}{\text{hour}^2}$ it must be multiplied by 0.464 876. Furthermore $\frac{dp}{p}$ may be replaced by 2.302 59 $d\ (\log p)$ by introducing Briggsian instead of natural logarithms and we then write (17) in the form

$$E_{p_0}^{p_1} = 1\ 837.3 \int_{p_1}^{p_0} (t + 459.4) d(\log p) \tag{18}$$

where t indicates degrees Fahrenheit, but p may be of any system of units since $d(\log p)$ is non-dimensional.

By treating equation (16) in a similar way we obtain

$$\Pi_{v_0}^{v_1} = p_{v_0}(1 - 10^{-\frac{1}{1837.3}\int_{v_0}^{v_1} \frac{dV}{t+459.4}}). \tag{19}$$

Moist air has a somewhat greater specific volume than dry air at the same temperature and pressure; but by applying an appropriate correction to the temperature, the Mariotte-Gay-Lussac law and formulas (18) and (19) can be made applicable to moist air also. To determine this correction we start with the equation of condition for moist air, viz. :

$$\frac{v(p - 0.377rf)}{T} = \frac{p_0 v_0}{T_0}.$$

where $r =$ relative humidity and $f =$ tension of saturated water-vapor at the temperature T. We have now to apply such a correction to T that the equation may be written in the Mariotte-Gay-Lussac form and yet give a true value of v. We therefore write

$$p \cdot \frac{v}{T_r} = p_0 \cdot \frac{v_0}{T_0}$$

where T_r expresses the corrected temperature. By eliminating v from these last two equations it is found that

$$T_r = \frac{pT}{p - 0.377r \cdot f}.$$

By subtracting T from both members this gives the correction

$$T_r - T = \frac{0.377r \cdot f \cdot T}{p - 0.377r \cdot f},$$

which by translating the above temperatures from the absolute to the Fahrenheit scale, may be written

$$t_r - t = \frac{0.377r \cdot f \cdot (t + 459.4)}{p - 0.377r \cdot f}, \tag{20}$$

where t_r is the " virtual temperature " of Guldberg and Mohn on the Fahrenheit scale. For purposes of tabulation we make $r = 1$ in equation (20), thus obtaining as the correction for saturated air

$$t_1 - t = \frac{0.377f \cdot (t + 459.4)}{p - 0.377f}.$$

Table 7 gives $t_1 - t$ for each inch of the mercurial barometer and each Fahrenheit degree.   In order to derive $t_r - t$ from $t_1 - t$ and r, the approximate formula

$$t_r - t = r(t_1 - t)$$

suffices. Table 8 gives $t_r - t$ for each five per cent. of relative humidity and each half degree of the quantity $t_1 - t$.

TABLE 7.

THE VALUES OF $t_1 - t$.

t Temp. °F.	19.0	20.0	21.0	22.0	23 0	24.0	25.0	26.0	27.0	28.0	29.0	30.0	31.0	t Temp. °F.
					$p =$ Pressure in Inches.									
0	0.5	0.5	0.5	0.5	0.5	0.5	0.5	0.5	0.5	0.5	0.5	0.5	0.5	0
10	0.5	0.5	0 5	0.5	0.5	0.5	0.5	0.5	0.5	0.5	0.5	0.5	0.5	10
20	1.0	1.0	1.0	1.0	1.0	1.0	1.0	1.0	0.5	0.5	0.5	0.5	0.5	20
30	1.5	1.5	1.5	1.5	1.5	1.5	1.0	1.0	1.0	1.0	1.0	1.0	1.0	30
35	2.0	2.0	2.0	2.0	1.5	1.5	1.5	1.5	1.5	1.5	1.5	1.5	1.5	35
40	2.5	2.5	2.0	2.0	2.0	2.0	2.0	2.0	1.5	1.5	1.5	1.5	1.5	40
45	3.0	3.0	2.5	2.5	2.5	2.5	2.5	2.0	2.0	2.0	2.0	2.0	2.0	45
50	3.5	3.5	3.5	3.0	3.0	3.0	3.0	2.5	2.5	2.5	2.5	2.5	2.0	50
51	4.0	3.5	3.5	3.5	3.0	3.0	3.0	3.0	2.5	2.5	2.5	2.5	2.5	51
52	4.0	3.5	3.5	3.5	3.5	3.0	3.0	3.0	2.0	2.5	2.5	2.5	2.5	52
53	4.0	4.0	3.5	3.5	3.5	3.5	3.0	3.0	3.0	3.0	2.5	2.5	2.5	53
54	4.0	4.0	4.0	3.5	3.5	3.5	3.5	3.0	3.0	3.0	3.0	2.5	2.5	54
55	4.5	4.0	4.0	4.0	3.5	3.5	3.5	3.0	3.0	3.0	3.0	3.0	2.5	55
56	4.5	4.5	4.0	4.0	4.0	4.0	3.5	3.5	3.0	3.0	3.0	3.0	3.0	56
57	4.5	4.5	4.5	4.0	4.0	4.0	4.0	3.5	3.5	3.0	3.0	3.0	3.0	57
58	5.0	4.5	4.5	4.5	4.0	4.0	4.0	4.0	3.5	3.5	3.5	3.0	3.0	58
59	5.0	5.0	4.5	4.5	4.5	4.0	4.0	4.0	3.5	3.5	3.5	3.5	3.0	59
60	5.5	5.0	5.0	4.5	4.5	4.5	4.0	4.0	3.5	3.5	3.5	3.5	3.5	60
61	5.5	5.0	5.0	5.0	4.5	4.5	4.5	4.0	4.0	4.0	3.5	3.5	3.5	61
62	6.0	5.5	5.0	5.0	5.0	5.0	4.5	4.0	4.0	4.0	4.0	3.5	3.5	62
63	6.0	5.5	5.5	5.0	5.0	5.0	4.5	4.5	4.0	4.0	4.0	4.0	3.5	63
64	6.0	6.0	5.5	5.5	5.0	5.0	4.5	4.5	4.5	4.0	4.0	4.0	4.0	64
65	6.5	6.0	6.0	5.5	5.5	5.0	5.0	4.5	4.5	4.5	4.0	4.0	4.0	65
66	6.5	6.5	6.0	6.0	5.5	5.5	5.0	5.0	4.5	4.5	4.5	4.0	4.0	66
67	7.0	6.5	6.5	6.0	6.0	5.5	5.5	5.0	5.0	4.5	4.5	4.5	4.0	67
68	7.0	7.0	6.5	6.5	6.0	5.5	5.5	5.5	5.0	5.0	4.5	4.5	4.5	68
69	7.5	7.0	7.0	6.5	6.0	6.0	5.5	5.5	5.5	5.0	5.0	5.0	4.5	69
70	7.5	7.5	7.0	6.5	6.5	6.0	6.0	5.5	5.5	5.5	5.0	5.0	5.0	70
71	8.0	7.5	7.5	7.0	6.5	6.5	6.0	6.0	5.5	5.5	5.5	5.0	5.0	71
72	8.5	8.0	7.5	7.0	7.0	6.5	6.5	6.0	6.0	5.5	5.5	5.5	5.0	72
73	8.5	8.5	8.0	7.5	7.0	7.0	6.5	6.5	6.0	6.0	5.5	5.5	5.5	73
74	9.0	8.5	8.0	8.0	7.5	7.0	7.0	6.5	6.5	6.0	6.0	5.5	5.5	74
75	—	9.0	8.5	8.0	7.5	7.5	7.0	7.0	6.5	6.5	6.0	6.0	5.5	75
76	—	9.0	8.5	8.5	8.0	7.5	7.5	7.0	7.0	6.5	6.5	6.0	6.0	76
77	—	9.5	9.0	8.5	8.5	8.0	7.5	7.5	7.0	7.0	6.5	6.5	6.0	77
78	—	10.0	9.5	9.0	8.5	8.0	8.0	7.5	7.5	7.0	7.0	6.5	6.5	78
79	—	10.0	9.5	9.5	9.0	8.5	8.0	8.0	7.5	7.5	7.0	7.0	6.5	79
80	—	—	—	9.5	9.0	9.0	8.5	8.0	8.0	7.5	7.5	7.0	7.0	80
81	—	—	—	10.0	9.5	9.0	9.0	8.5	8.0	8.0	7.5	7.5	7.0	81
82	—	—	—	10.5	10.0	9.5	9.0	8.5	8.5	8.0	8.0	7.5	7.5	82
83	—	—	—	10.5	10.0	10.0	9.5	9.0	8.5	8.5	8.0	8.0	7.5	83
84	—	—	—	11.0	10.5	10.0	9.5	9.5	9.0	8.5	8.5	8.0	8.0	84
85	—	—	—	—	—	10.5	10.0	9.5	9.5	9.0	8.5	8.5	8.0	85
86	—	—	—	—	—	11.0	10.5	10.0	9.5	9.5	9.0	8.5	8.5	86
87	—	—	—	—	—	11.0	11.0	10.5	10.0	9.5	9.0	9.0	8.5	87
88	—	—	—	—	—	11.5	11.0	10.5	10.5	10.0	9.5	9.0	9.0	88
89	—	—	—	—	—	12.0	11.5	11.0	10.5	10.0	10.0	9.5	9.0	89

TABLE 7 (*Continued*).

THE VALUES OF t_1-t.

t Temp. °F.	\multicolumn{13}{c}{$p =$ Pressure in Inches.}	t Temp. °F.
	19.0	20.0	21.0	22.0	23.0	24.0	25 0	26.0	27.0	28.0	29.0	30.0	31.0	
90	—	—	—	—	—	—	—	11.5	11.0	10.5	10.0	10.0	9.5	90
91	—	—	—	—	—	—	—	12.0	11.5	11.0	10.5	10.5	10.0	91
92	—	—	—	—	—	—	—	12.5	12.0	11.5	11.0	10.5	10.5	92
93	—	—	—	—	—	—	—	12.5	12.0	12.0	11.5	11.0	10.5	93
94	—	—	—	—	—	—	—	13.0	12.5	12.0	11.5	11.5	11.0	94
95	—	—	—	—	—	—	—	—	—	12.5	12.0	11.5	11.5	95
96	—	—	—	—	—	—	—	—	—	13.0	12.5	12.0	11.5	96
97	—	—	—	—	—	—	—	—	—	13.5	13.0	12.5	12.0	97
98	—	—	—	—	—	—	—	—	—	14.0	13.5	13.0	12.5	98
99	—	—	—	—	—	—	—	—	—	14.5	14.0	13.5	13.0	99

TABLE 8.

THE VALUES OF t_1-t.

t_1-t ° Fahr.	\multicolumn{21}{c}{Percentage of Relative Humidity.}	t_1-t ° Fahr.
	0	5	10	15	20	25	30	35	40	45	50	55	60	65	70	75	80	85	90	95	100	
0.5	0.0	0.0	0.0	0.0	0.0	0.0	0.0	0.0	0.0	0.0	0.5	0.5	0.5	0.5	0.5	0.5	0.5	0.5	0.5	0.5	0 5	0.5
1.0	0.0	0.0	0.0	0.0	0.0	0.5	0.5	0.5	0.5	0.5	0.5	0.5	0.5	0.5	1.0	1.0	1.0	1.0	1.0	1.0	1.0	1.0
1.5	0 0	0.0	0.0	0.0	0.5	0.5	0.5	0.5	0.5	0.5	1.0	1.0	1.0	1.0	1 0	1 0	1.5	1.5	1.5	1 5	1.5	1.5
2.0	0.0	0.0	0.0	0.5	0.5	0.5	0.5	0.5	1.0	1.0	1.0	1.0	1.0	1.5	1.5	1.5	1.5	1.5	2.0	2.0	2 0	2.0
2.5	0.0	0.0	0.5	0.5	0.5	0.5	1.0	1.0	1.0	1.5	1.5	1.5	1.5	2.0	2.0	2.0	2.0	2.5	2.5	2.5	2.5	2.5
3.0	0.0	0.0	0.5	0.5	0.5	1.0	1.0	1.0	1.5	1.5	1.5	2.0	2.0	2.0	2.0	2.5	2.5	2.5	3.0	3.0	3 0	3 0
3.5	0.0	0.0	0.5	0.5	0.5	1.0	1.0	1.0	1.5	1.5	2.0	2.0	2.0	2 5	2.5	2.5	3.0	3.0	3.0	3 5	3.5	3 5
4.0	0.0	0.0	0.5	0 5	1.0	1.0	1.0	1.5	1.5	2.0	2.0	2.0	2.5	2.5	3.0	3.0	3.0	3.5	3.5	4.0	4.0	4 0
4.5	0.0	0.0	0.5	0.5	1.0	1.0	1.5	1.5	2.0	2.0	2.5	2.5	2.5	3.0	3.0	3.5	3.5	4.0	4.0	4.5	4.5	4.5
5.0	0.0	0.5	0.5	1.0	1.0	1.5	1.5	2.0	2.0	2.5	2.5	3.0	3.0	3.5	3 5	4.0	4.0	4.5	4.5	5.0	5.0	5.0
5.5	0.0	0.5	0.5	1.0	1.0	1.5	1.5	2.0	2.0	2.5	3.0	3.0	3.5	3.5	4.0	4.0	4.5	4.5	5.0	5.0	5.5	5.5
6.0	0.0	0.5	0.5	1 0	1.0	1.5	2.0	2.0	2.5	2.5	3.0	3.5	3.5	4.0	4.0	4.5	5.0	5.0	5.5	5.5	6 0	6.0
6.5	0.0	0.5	0.5	1.0	1.5	1.5	2.0	2.5	2.5	3.0	3.5	3.5	4.0	4.0	4.5	5.0	5 0	5.5	6.0	6.0	6.5	6.5
7.0	0.0	0.5	0.5	1.0	1.5	2.0	2.0	2.5	3.0	3.0	3.5	4.0	4.0	4.5	5.0	5.5	5.5	6.0	6.5	6.5	7.0	7.0
7.5	0.0	0.5	1.0	1.0	1.5	2.0	2.5	2.5	3.0	3.5	4.0	4.0	4.5	5.0	5.5	5.5	6.0	6.5	7.0	7.0	7.5	7.5
8.0	0.0	0.5	1.0	1.0	1.5	2.0	2.5	3.0	3.0	3.5	4.0	4.5	5.0	5.0	5.5	6.0	6.5	7.0	7.0	7.5	8.0	8.0
8.5	0.0	0.5	1.0	1.5	1.5	2.0	2.5	3.0	3.5	4.0	4.5	4.5	5.0	5.5	6.0	6.5	7.0	7.0	7.5	8.0	8.5	8 5
9.0	0.0	0.5	1.0	1.5	2.0	2.5	2.5	3.0	3.5	4.0	4.5	5.0	5.5	6.0	6.5	7.0	7.0	7.5	8.0	8.5	9.0	9.0
9.5	0.0	0.5	1.0	1.5	2.0	2.5	3.0	3.5	4.0	4.5	5.0	5.0	5.5	6.0	6.5	7.0	7.5	8.0	8.5	9.0	9.5	9.5
10.0	0.0	0.5	1.0	1.5	2.0	2.5	3.0	3.5	4.0	4.5	5.0	5.5	6.0	6.5	7.0	7.5	8.0	8.5	9.0	9.5	10.0	10.0
10.5	0.0	0.5	1.0	1.5	2.0	2.5	3.0	3.5	4.0	4.5	5.5	6.0	6.5	7.0	7.5	8.0	8.5	9.0	9.5	10.0	10.5	10.5
11.0	0.0	0.5	1.0	1.5	2.0	3.0	3.5	4.0	4.5	5.0	5.5	6.0	6.5	7.0	7.5	8.5	9.0	9.5	10.0	10.5	11 0	11 0
11.5	0.0	0.5	1.0	1.5	2.5	3.0	3.5	4.0	4.5	5 0	6.0	6.5	7.0	7.5	8.0	8 5	9.5	10.0	10 5	11.0	11 5	11.5
12.0	0.0	0.5	1.0	2.0	2.5	3 0	3.5	4.0	5.0	5.5	6 0	6.5	7.0	8 0	8.5	9.0	9.5	10.0	11.0	11.5	12.0	12.0
12.5	0.0	0.5	1.5	2.0	2.5	3.0	4.0	4.5	5.0	5.5	7.0	7.5	8.0	9.0	9.5	10.0	10 5	11.5	12.0	12.5	12.5
13.0	0.0	0.5	1.5	2.0	2.5	3.5	4.0	4.5	5.0	6.0	6.5	7.0	8.0	8.5	9.0	10.0	10.5	11.0	11.5	12.5	13 0	13.0
13.5	0.0	0.5	1.5	2.0	2.5	3.5	4.0	4.5	5.5	6.0	7.0	7.5	8.0	9.0	9.5	10.0	11.0	11.5	12.0	13.0	13 5	13.5
14.0	0.0	0 5	1.5	2.0	3.0	3.5	4.0	5.0	5.5	6.5	7.0	7.5	8.5	9.0	10.0	10.5	11.0	11.5	12.5	13.0	14.0	14.0

Example: During a kite ascension made at Omaha on Sept. 23, 1898, at 11.25 A. M., 75th meridian standard time, there was observed $p = 24.20$ inches, $t = 68°$ F., $r = 51$ per cent.

Table 7, for $p = 24.20$ inches and $t = 68°$ F., gives $t_1 - t = 5°.5$; and

Table 8, for $t_1 - t = 5°.5$ and $r = 51$ per cent., gives $t_r - t = 3°.0$. The virtual temperature is thus found to be $68° + 3° = 71°$ F.

Formulæ (18) and (19) can be made valid for moist air if t_r be substituted for t in them, and they then read

$$E_{p_0}^{p_1} = 1\,837.3 \int_{p_1}^{p_0} (t_r + 459.4) d(\log p), \tag{21}$$

$$\Pi_{V_0}^{V_1} = p_{V_0}(1 - 10^{-\frac{1}{1837.3} \int_{V_0}^{V_1} \frac{dV}{t_r + 459.4}}). \tag{22}$$

The condition for atmospheric equilibrium may be so formulated that the number $\Pi_{V_0}^{V}$ of isobaric surfaces contained between two level surfaces, $V = V_0$ and $V = V_1$ is everywhere the same. From equation (22) it appears that this is the case when t_r can be expressed as a function of V alone, *i. e.*, when the surfaces of equal values of t_r coincide with the level surfaces of gravity. Whence it appears that in an atmosphere in the condition of static equilibrium the surfaces of equal values of t_r, as well as the isobaric surfaces, coincide with the level surfaces of gravity.

The values of $E_{p_0}^{p_1}$ and of $\Pi_{V_0}^{V}$ may be easily tabulated if we restrict ourselves once for all to a small number of limiting values of p_0 and p_1 as well as of V_0 and V_1. For example, we choose respectively every half-inch of barometric pressure and every 10 000th level surface of gravity, that is to say we compute the following values:

$$E_{30.0}^{29.5} \quad E_{29.5}^{29.0} \quad E_{29.0}^{28.5} \quad E_{28.5}^{28.0} \quad E_{28.0}^{27.5} \text{ etc.,}$$

$$\Pi_0^{10\,000} \quad \Pi_{10\,000}^{20\,000} \quad \Pi_{20\,000}^{30\,000} \quad \Pi_{30\,000}^{40\,000} \text{ etc.}$$

For such small intervals the average values of t_r may be readily found by graphic interpolation. When these values are substituted in (21) and (22) and the latter are then integrated we obtain:

$$E_{p_0}^{p_1} = 1\,837.3(t_r + 459.4)\frac{\log p_0}{p_1} \tag{23}$$

and

$$\Pi_{V_0}^{V_1} = p_{V_0}(1 - 10^{-\frac{V_1 - V_0}{1837.3(t_r + 459.4)}}). \tag{24}$$

From equation (23) are obtained the following :

$E_{31.0}^{30.5} = 12.966(t_r + 459.4)$ $\quad E_{27.0}^{26.5} = 14.920(t_r + 559.4)$ $\quad E_{23.0}^{22.3} = 17.535(t_r + 459.4)$

$E_{30.5}^{30.0} = 13.186(t_r + 459.4)$ $\quad E_{26.5}^{26.0} = 15.206(t_r + 459.4)$ $\quad E_{22.5}^{22.0} = 17.929(t_r + 459.4)$

$E_{30.0}^{29.5} = 13.410(t_r + 459.4)$ $\quad E_{26.0}^{25.5} = 15.498(t_r + 459.4)$ $\quad E_{22.0}^{21.5} = 18.340(t_r + 459.4)$

$E_{29.5}^{29.0} = 13.640(t_r + 459.4)$ $\quad E_{25.5}^{25.0} = 15.801(t_r + 459.4)$ $\quad E_{21.5}^{21.0} = 18.773(t_r + 459.4)$

$E_{29.0}^{28.5} = 13.877(t_r + 459.4)$ $\quad E_{2.50}^{2.45} = 16.116(t_r + 459.4)$ $\quad E_{21.0}^{20.5} = 19.230(t_r + 459.4)$

$E_{28.5}^{28.0} = 14.122(t_r + 459.4)$ $\quad E_{24.5}^{24.0} = 16.445(t_r + 459.4)$ $\quad E_{20.5}^{20.0} = 19.703(t_r + 459.4)$

$E_{28.0}^{27.5} = 14.375(t_r + 459.4)$ $\quad E_{24.0}^{23.5} = 16.788(t_r + 459.4)$ $\quad E_{20.0}^{19.5} = 20.204(t_r + 459.4)$

$E_{27.5}^{27.0} = 14.640(t_r + 459.4)$ $\quad E_{23.5}^{23.0} = 17.148(t_r + 459.4)$ $\quad E_{19.5}^{19.0} = 20.736(t_r + 459.4)$

From equation (24) there results

$$\Pi_V^{V+10\,000} = p_V \left\{1 - 10^{-\frac{10\,000}{1837.3(t_r + 459.4)}}\right\}.$$

Table 9 contains the values of $E_{31.0}^{30.5} \cdots E_{19.5}^{19.0}$ for each whole degree Fahrenheit of the virtual temperature between the limits $t_r = 15°$ and $t_r = 99°$.

Table 10 contains the values of $\Pi_V^{V+10\,000}$ as a function of p_V and t_r for every tenth of an inch of barometric pressure between the limits $p_V = 19.0$ inches and $p_V = 30.9$ inches and for every ten degrees of the Fahrenheit scale.

In calculating the value of p_V those level surfaces of gravity that lie beneath the surface of the earth are of course to be excluded. We compute first the pressure for the first level surface above the ground that is a whole multiple of 10 000. For example, in Omaha this would be $V = 20\,000$ since the station-barometer there is in the level surface 18 550. If we substitute $V_0 = 18\,550$ and $V_1 = 20\,000$ in (24) we obtain the difference in pressure between the level surface of gravity $V = 20\,000$ and the station-barometer at Omaha, viz.:

$$\Pi_{18\,550}^{20\,000} = p_{18\,550} - p_{20\,000},$$

$$= p_{18\,550} \left\{1 - 10^{-\frac{1\,450}{1837.3(t_r + 459.4)}}\right\},$$

Table 11 contains these values of $\Pi_{18\,550}^{20\,000}$ expressed as a function of the pressure $p_{18\,550}$ recorded by the station-barometer at Omaha, and the mean virtual temperature, t_r, between $V = 18\,550$ and $V = 20\,000$.

TABLE 9.

Temp. °F	E19.0/19.5	E19.5/20.0	E20.0/20.5	E20.5/21.0	E21.0/21.5	E21.5/22.0	E22.0/22.5	E22.5/23.0	E23.0/23.5	E23.5/24.0	E24.0/24.5	E24.5/25.0	E25.0/25.5	E25.5/26.0	E26.0/26.5	E26.5/27.0	E27.0/27.5	E27.5/28.0	E28.0/28.5	E28.5/29.0	E29.0/29.5	E29.5/30.0	E30.0/30.5	E30.5/31.0
15	9830	9580	9350	9120	8910	8700	8510	8320	8140	7970	7800	7650	7500	7350	7210	7080	6950	6820	6700	6580	6470	6360	6260	6150
16	9860	9600	9370	9140	8930	8720	8530	8340	8160	7990	7820	7660	7510	7360	7230	7090	6960	6830	6720	6600	6480	6380	6270	6170
17	9880	9620	9390	9160	8940	8740	8540	8350	8180	8000	7840	7680	7530	7390	7240	7110	6980	6840	6730	6610	6500	6390	6280	6180
18	9900	9640	9410	9180	8960	8760	8560	8370	8190	8020	7850	7700	7540	7400	7260	7120	6990	6860	6740	6620	6510	6400	6290	6190
19	9920	9660	9430	9200	8980	8780	8580	8390	8210	8040	7870	7710	7560	7410	7270	7140	7010	6870	6760	6640	6520	6420	6310	6210
20	9940	9680	9450	9220	9000	8800	8600	8410	8230	8050	7890	7730	7570	7430	7290	7150	7020	6890	6770	6650	6540	6430	6320	6220
21	9960	9700	9470	9240	9020	8810	8620	8420	8240	8070	7900	7750	7590	7440	7300	7170	7030	6900	6790	6660	6550	6440	6340	6230
22	9980	9720	9490	9260	9040	8830	8630	8440	8260	8090	7920	7760	7610	7460	7320	7180	7050	6920	6800	6680	6570	6460	6350	6240
23	10000	9740	9500	9280	9060	8850	8650	8460	8280	8110	7930	7780	7620	7470	7330	7200	7060	6930	6810	6690	6580	6470	6360	6260
24	10020	9770	9530	9290	9080	8870	8670	8480	8300	8120	7950	7790	7640	7490	7350	7210	7080	6950	6830	6710	6590	6480	6380	6270
25	10040	9790	9540	9310	9090	8890	8690	8490	8310	8140	7970	7810	7650	7500	7360	7230	7090	6960	6840	6720	6610	6500	6390	6280
26	10060	9810	9560	9330	9110	8910	8710	8510	8330	8160	7990	7830	7670	7520	7380	7240	7110	6970	6860	6730	6620	6510	6400	6300
27	10080	9830	9580	9350	9130	8920	8720	8530	8350	8180	8000	7840	7690	7540	7400	7260	7120	6990	6870	6750	6630	6520	6420	6310
28	10100	9850	9610	9370	9150	8940	8740	8550	8360	8190	8020	7860	7700	7550	7410	7270	7140	7000	6880	6760	6650	6540	6430	6320
29	10120	9870	9620	9390	9170	8960	8760	8560	8380	8210	8030	7870	7720	7570	7430	7290	7150	7020	6900	6780	6660	6550	6440	6340
30	10150	9890	9640	9410	9190	8980	8780	8580	8400	8230	8050	7890	7730	7580	7440	7300	7170	7030	6910	6790	6670	6570	6460	6350
31	10170	9910	9660	9430	9210	9000	8800	8600	8420	8240	8070	7910	7750	7600	7450	7320	7180	7050	6930	6800	6690	6580	6470	6360
32	10190	9930	9680	9450	9230	9020	8810	8620	8430	8260	8080	7920	7760	7610	7470	7330	7200	7060	6940	6820	6700	6590	6480	6370
33	10210	9950	9700	9470	9240	9030	8830	8630	8450	8270	8100	7940	7780	7630	7480	7350	7210	7070	6960	6830	6720	6610	6500	6390
34	10230	9970	9720	9490	9260	9050	8850	8650	8470	8290	8120	7950	7800	7640	7500	7360	7220	7090	6970	6840	6730	6620	6511	6400
35	10250	9990	9740	9510	9280	9070	8870	8670	8480	8310	8130	7970	7810	7660	7520	7370	7240	7100	6980	6860	6740	6630	6520	6410
36	10270	10010	9760	9530	9300	9090	8880	8690	8500	8320	8150	7990	7830	7670	7520	7390	7250	7120	7000	6870	6760	6650	6540	6430
37	10290	10030	9780	9540	9320	9110	8900	8700	8520	8340	8170	8000	7840	7690	7540	7400	7270	7130	7010	6890	6770	6660	6550	6440
38	10310	10050	9800	9560	9340	9130	8920	8720	8540	8360	8180	8020	7860	7710	7560	7410	7280	7150	7030	6900	6780	6670	6560	6450
39	10330	10070	9820	9580	9360	9140	8940	8740	8550	8370	8200	8040	7880	7720	7570	7430	7300	7160	7040	6910	6800	6680	6570	6460
40	10350	10090	9840	9600	9380	9160	8960	8760	8570	8390	8210	8050	7890	7740	7590	7450	7310	7180	7050	6930	6810	6700	6590	6480
41	10370	10110	9860	9620	9400	9180	8970	8770	8590	8410	8230	8070	7910	7750	7600	7460	7330	7190	7070	6940	6820	6710	6600	6490
42	10390	10130	9880	9640	9410	9200	8990	8790	8600	8420	8250	8080	7920	7770	7620	7480	7340	7200	7080	6960	6840	6720	6610	6500
43	10410	10150	9900	9660	9430	9220	9010	8810	8620	8440	8260	8100	7940	7780	7630	7490	7360	7220	7100	6970	6850	6730	6630	6520
44	10440	10170	9920	9680	9450	9240	9030	8830	8640	8460	8280	8120	7950	7800	7650	7510	7370	7230	7110	6980	6870	6750	6640	6530
45	10460	10190	9940	9700	9470	9250	9050	8840	8660	8470	8300	8130	7970	7810	7660	7520	7390	7250	7120	7000	6880	6760	6650	6540
46	10480	10210	9960	9720	9490	9270	9060	8860	8670	8490	8310	8150	7980	7830	7680	7540	7400	7260	7140	7010	6890	6780	6670	6560
47	10500	10230	9980	9740	9510	9290	9080	8880	8690	8510	8330	8160	8000	7850	7700	7550	7420	7280	7160	7030	6910	6790	6680	6570
48	10520	10250	10000	9760	9530	9310	9100	8900	8710	8520	8350	8180	8020	7860	7700	7570	7430	7290	7170	7040	6920	6800	6690	6580
49	10540	10270	10020	9780	9550	9330	9120	8910	8720	8540	8360	8200	8030	7880	7720	7580	7440	7300	7180	7050	6930	6810	6710	6590
50	10560	10290	10040	9790	9560	9350	9140	8930	8740	8560	8380	8210	8050	7890	7740	7600	7460	7320	7200	7070	6950	6830	6720	6610
51	10580	10310	10060	9810	9580	9360	9150	8950	8760	8580	8400	8230	8060	7910	7760	7610	7470	7330	7210	7080	6960	6840	6730	6620
52	10600	10330	10080	9830	9600	9380	9170	8970	8780	8600	8410	8250	8080	7920	7770	7630	7490	7350	7220	7090	6970	6850	6750	6630
53	10620	10350	10100	9850	9620	9400	9190	8980	8790	8610	8430	8260	8100	7940	7790	7640	7500	7360	7240	7110	6990	6860	6760	6650
54	10640	10370	10120	9870	9640	9420	9210	9000	8810	8630	8440	8280	8110	7950	7800	7660	7520	7380	7250	7120	7000	6880	6770	6660
55	—	10390	10140	9890	9660	9440	9230	9020	8830	8640	8460	8290	8130	7970	7820	7670	7530	7390	7270	7140	7020	6900	6790	6670
56	—	10410	10150	9910	9680	9460	9240	9040	8840	8660	8480	8310	8140	7980	7830	7690	7550	7410	7290	7150	7030	6910	6800	6690
57	—	—	10170	9930	9700	9470	9260	9060	8860	8680	8490	8330	8160	8000	7850	7700	7560	7420	7300	7160	7040	6920	6810	6700
58	—	—	10190	9950	9710	9490	9280	9070	8880	8690	8510	8340	8180	8020	7860	7720	7580	7430	7310	7180	7060	6940	6830	6710
59	—	—	10210	9970	9730	9510	9300	9090	8900	8710	8530	8360	8190	8030	7880	7730	7590	7450	7320	7190	7070	6960	6840	6720

TABLE 9 (*Concluded*).

t Temp. °F	$E^{19.0}_{19.5}$	$E^{19.5}_{20.0}$	$E^{20.0}_{20.5}$	$E^{20.5}_{21.0}$	$E^{21.0}_{21.5}$	$E^{21.5}_{22.0}$	$E^{22.0}_{22.5}$	$E^{22.5}_{23.0}$	$E^{23.0}_{23.5}$	$E^{23.5}_{24.0}$	$E^{24.0}_{24.5}$	$E^{24.5}_{25.0}$	$E^{25.0}_{25.5}$	$E^{25.5}_{26.0}$	$E^{26.0}_{26.5}$	$E^{26.5}_{27.0}$	$E^{27.0}_{27.5}$	$E^{27.5}_{28.0}$	$E^{28.0}_{28.5}$	$E^{28.5}_{29.0}$	$E^{29.0}_{29.5}$	$E^{29.5}_{30.0}$	$E^{30.0}_{30.5}$	$E^{30.5}_{31.0}$
60	—	—	—	9990	9750	9530	9310	9110	8910	8730	8540	8370	8210	8050	7890	7750	7610	7460	7340	7210	7080	6970	6850	6740
61	—	—	—	10010	9770	9550	9330	9130	8930	8740	8560	8390	8220	8060	7910	7760	7620	7480	7350	7220	7100	6980	6870	6750
62	—	—	—	—	9790	9570	9350	9140	8950	8760	8580	8400	8240	8080	7920	7780	7630	7490	7360	7230	7110	6990	6880	6760
63	—	—	—	—	9810	9580	9370	9160	8960	8780	8590	8420	8250	8090	7940	7790	7650	7510	7380	7250	7120	7010	6890	6780
64	—	—	—	—	9830	9600	9390	9180	8980	8790	8610	8440	8270	8110	7950	7810	7660	7520	7390	7260	7140	7020	6900	6790
65	—	—	—	—	—	9620	9400	9200	9000	8810	8630	8450	8290	8120	7970	7820	7680	7530	7410	7280	7150	7030	6920	6800
66	—	—	—	—	—	9640	9420	9210	9020	8830	8640	8470	8300	8140	7990	7840	7690	7550	7420	7290	7170	7050	6930	6810
67	—	—	—	—	—	—	9440	9230	9030	8840	8660	8490	8320	8150	8000	7850	7710	7560	7440	7300	7180	7060	6940	6830
68	—	—	—	—	—	—	9460	9250	9050	8860	8680	8500	8330	8170	8020	7870	7720	7580	7450	7320	7190	7070	6960	6840
69	—	—	—	—	—	—	9480	9270	9070	8880	8690	8520	8350	8190	8030	7890	7740	7590	7460	7330	7210	7090	6970	6850
70	—	—	—	—	—	—	—	9280	9080	8890	8710	8530	8370	8200	8050	7900	7750	7610	7480	7340	7220	7110	6980	6870
71	—	—	—	—	—	—	—	9300	9100	8910	8720	8550	8380	8220	8060	7910	7770	7620	7490	7360	7230	7120	7000	6880
72	—	—	—	—	—	—	—	—	9120	8930	8740	8570	8400	8230	8080	7930	7780	7640	7510	7370	7250	7130	7010	6890
73	—	—	—	—	—	—	—	—	9140	8950	8760	8580	8410	8250	8090	7940	7800	7650	7520	7390	7260	7140	7020	6910
74	—	—	—	—	—	—	—	—	9150	8960	8770	8600	8430	8260	8110	7960	7810	7660	7530	7400	7270	7150	7040	6920
75	—	—	—	—	—	—	—	—	—	8980	8790	8620	8440	8280	8120	7970	7830	7680	7550	7410	7290	7170	7050	6930
76	—	—	—	—	—	—	—	—	—	9000	8810	8630	8460	8290	8140	7990	7840	7690	7560	7430	7300	7180	7060	6940
77	—	—	—	—	—	—	—	—	—	—	8820	8650	8480	8310	8150	8000	7850	7710	7580	7440	7320	7200	7080	6960
78	—	—	—	—	—	—	—	—	—	—	8840	8660	8490	8330	8170	8020	7860	7720	7590	7460	7330	7210	7090	6970
79	—	—	—	—	—	—	—	—	—	—	8860	8680	8510	8340	8180	8030	7880	7740	7600	7470	7340	7220	7100	6980
80	—	—	—	—	—	—	—	—	—	—	—	8700	8520	8360	8200	8050	7900	7750	7620	7480	7360	7240	7120	7000
81	—	—	—	—	—	—	—	—	—	—	—	8710	8540	8370	8210	8060	7910	7760	7630	7500	7370	7250	7130	7010
82	—	—	—	—	—	—	—	—	—	—	—	—	8550	8390	8230	8080	7930	7780	7650	7510	7380	7260	7140	7020
83	—	—	—	—	—	—	—	—	—	—	—	—	8570	8400	8240	8090	7940	7790	7660	7520	7400	7280	7160	7040
84	—	—	—	—	—	—	—	—	—	—	—	—	8590	8420	8260	8110	7960	7810	7680	7540	7410	7290	7170	7050
85	—	—	—	—	—	—	—	—	—	—	—	—	—	8430	8270	8120	7970	7820	7690	7550	7420	7300	7180	7060
86	—	—	—	—	—	—	—	—	—	—	—	—	—	8450	8290	8140	7990	7840	7700	7570	7440	7310	7190	7070
87	—	—	—	—	—	—	—	—	—	—	—	—	—	—	8300	8150	8000	7850	7720	7580	7450	7330	7210	7080
88	—	—	—	—	—	—	—	—	—	—	—	—	—	—	8320	8170	8020	7870	7730	7590	7470	7340	7220	7100
89	—	—	—	—	—	—	—	—	—	—	—	—	—	—	8330	8180	8030	7880	7750	7610	7480	7350	7230	7110
90	—	—	—	—	—	—	—	—	—	—	—	—	—	—	—	8200	8040	7890	7760	7620	7490	7370	7250	7130
91	—	—	—	—	—	—	—	—	—	—	—	—	—	—	—	8210	8060	7910	7770	7640	7500	7380	7260	7140
92	—	—	—	—	—	—	—	—	—	—	—	—	—	—	—	—	8070	7920	7790	7650	7520	7390	7270	7150
93	—	—	—	—	—	—	—	—	—	—	—	—	—	—	—	—	8090	7940	7800	7660	7530	7410	7290	7170
94	—	—	—	—	—	—	—	—	—	—	—	—	—	—	—	—	8100	7950	7820	7680	7550	7420	7300	7180
95	—	—	—	—	—	—	—	—	—	—	—	—	—	—	—	—	—	7970	7830	7690	7560	7430	7310	7190
96	—	—	—	—	—	—	—	—	—	—	—	—	—	—	—	—	—	7980	7840	7710	7570	7450	7330	7200
97	—	—	—	—	—	—	—	—	—	—	—	—	—	—	—	—	—	—	7860	7720	7590	7460	7340	7220
98	—	—	—	—	—	—	—	—	—	—	—	—	—	—	—	—	—	—	7870	7730	7600	7470	7350	7230
99	—	—	—	—	—	—	—	—	—	—	—	—	—	—	—	—	—	—	7890	7750	7620	7490	7370	7240

TABLE 10.

THE VALUES OF $\prod_r^{r+10\,000} = p_r - p_{r+10\,000}$

p_r	$t_r=0°$	10°	20°	30°	40°	50°	60°	70°	80°	90°	100°
19.0	0.511	0.501	0.490	0.480	0.471	0.462	0.453	0.444	·0.436	0.429	0.421
.1	.514	.503	.493	.483	.473	.464	.455	.447	.439	.431	.423
.2	.517	.506	.495	.485	.476	.467	.458	.449	.441	.433	.425
.3	.519	.509	.498	.488	.478	.469	.460	.451	.443	.435	.427
.4	.522	.511	.501	.490	.481	.471	.462	.454	.446	.438	.430
19.5	.525	.514	.503	.493	.483	.474	.465	.456	.448	.440	.432
.6	.527	.516	.506	.495	.486	.476	·.467	.458	.450	.442	.434
.7	.530	.519	.508	.498	.488	.479	.470	.461	.453	.444	.436
.8	.533	.522	.511	.501	.491	.481	.472	.463	.455	.447	.439
.9	.536	.524	.513	.503	.493	.484	.474	.465	.457	.449	.441
20.0	0.538	0.527	0.516	0.506	0.496	0.486	0.477	0.468	0.459	0.451	0.443
.1	.541	.530	.519	.508	.498	.488	.479	.470	.462	.453	.445
.2	.544	.532	.521	.511	.501	.491	.482	.472	.464	.456	.447
.3	.546	.535	.524	.513	.503	.493	.484	.475	.466	.458	.450
.4	.549	.538	.526	.516	.506	.496	.486	.477	.469	.460	.452
20.5	.552	.540	.529	.518	.508	.498	.489	.479	.471	.462	.454
.6	.554	.543	.531	.521	.510	.501	.491	.482	.473	.465	.456
.7	.557	.545	.534	.523	.513	.503	.493	.484	.475	.467	.459
.8	.560	.548	.537	.526	.515	.505	.496	.487	.478	.469	.461
.9	.562	.551	.539	.528	.518	.508	.498	.489	.480	.472	.463
21.0	0.565	0.553	0.542	0.531	0.520	0.510	0.501	0.491	0.482	0.474	0.465
.1	.568	·.556	.544	.533	.523	.513	.503	.494	.485	.476	.467
.2	·.570	.559	.547	.536	.525	.515	.505	.496	.487	.478	.470
.3	.573	.561	.550	.538	.528	.518	.508	.498	.489	.481	.472
.4	.576	.564	.552	.541	.530	.520	.510	.501	.492	.483	.474
21.5	.579	.567	.555	.544	.533	.522	.513	.503	.494	.485	.476
.6	.581	.569	.557	.546	.535	.525	.515	.505	.496	.487	.478
.7	.584	.572	.560	.549	.538	.527	.517	.508	.498	.490	.481
.8	.587	.574	.562	.551	.540	.530	.520	.510	.501	.492	.483
.9	.589	.577	.565	.554	.543	.532	.522	.512	.503	.494	.485
22.0	0.592	0.580	0.568	0.556	0.545	0.535	0.524	0.515	0.505	0.496	0.487
.1	.595	.582	.570	.559	.548	.537	.527	.517	.508	.499	.490
.2	.597	.585	.573	.561	.550	.539	.529	.519	.510	.501	.492
.3	.600	.588	.575	.564	.553	.542	.532	.522	.512	.503	.494
.4	.603	.590	·.578	.566	.555	.544	.534	.524	.515	.505	.496
22.5	.605	.593	.581	.569	.558	.547	.536	.526	.517	.508	.498
.6	.608	.596	.583	.571	.560	.549	.539	.529	.519	.510	.501
.7	.611	.598	.586	.574	.563	.552	.541	.531	.521	.512	.503
.8	.614	.601	.588	.576	.565	.554	.544	.533	.524	.514	.505
.9	.616	.603	.591	.579	.567	.556	.546	.536	.526	.517	.507
23.0	0.619	0.606	0.593	0.581	0.570	0.559	0.548	0.538	0.528	0.519	0.509
.1	.622	.609	.596	.584	.572	.561	.551	.540	.531	.521	.512
.2	.624	.611	.599	.586	.575	.564	.553	.543	.533	.523	.514
.3	.627	.614	.601	.589	.577	.566	.555	.545	.535	.526	.516
.4	.630	.617	.604	.592	.580	.569	.558	.547	.537	.528	.518
23.5	.632	.619	.606	.594	.582	.571	.560	.550	.540	.530	.521
.6	.635	.622	.609	.597	.585	.573	.563	.552	.542	.532	.523
.7	.638	.624	.611	.599	.587	.576	.565	.554	.544	.535	.525
.8	.640	.627	.614	.602	.590	.578	.567	.557	.547	.537	.527
.9	.643	.630	.617	.604	.592	.581	.570	.559	.549	.539	.529

TABLE 10.—*Continued.*

THE VALUES OF $\prod_\gamma^{\gamma+10\,000} = p_\gamma - p_{\gamma+10\,000}$.

p_γ	$t_\gamma=0°$	10°	20°	30°	40°	50°	60°	70°	80°	90°	100°
24.0	0.646	0.632	0.619	0.607	0.595	0.583	0.572	0.561	0.551	0.541	0.532
.1	.649	.635	.622	.609	.597	.586	.575	.564	.554	.544	.534
.2	.651	.638	.624	.612	.600	.588	.577	.566	.556	.546	.536
.3	.654	.640	.627	.614	.602	.590	.579	.568	.558	.548	.538
.4	.657	.643	.630	.617	.605	.593	.582	.571	.560	.550	.540
24.5	.659	.646	.632	.619	.607	.595	.584	.573	.563	.553	.543
.6	.662	.648	.635	.622	.610	.598	.586	.575	.565	.555	.545
.7	.665	.651	.637	.624	.612	.600	.589	.578	.567	.557	.547
.8	.667	.653	.640	.627	.615	.603	.591	.580	.570	.559	.549
.9	.670	.656	.642	.629	.617	.605	.594	.582	.572	.562	.552
25.0	0.673	0.659	0.645	0.632	0.620	0.608	0.596	0.585	0.574	0.564	0.554
.1	.675	.661	.648	.635	.622	.610	.598	.587	.577	.566	.556
.2	.678	.664	.650	.637	.624	.612	.601	.589	.579	.569	.558
.3	.681	.667	.653	.640	.627	.615	.603	.592	.581	.571	.560
.4	.684	.669	.655	.642	.629	.617	.606	.594	.583	.573	.563
25.5	.686	.672	.658	.645	.632	.620	.608	.596	.586	.575	.565
.6	.689	.675	.660	.647	.634	.622	.610	.599	.588	.578	.567
.7	.692	.677	.663	.650	.637	.625	.613	.601	.590	.580	.569
.8	.694	.680	.666	.652	.639	.627	.615	.603	.593	.582	.571
.9	.697	.682	.668	.655	.642	.629	.617	.606	.595	.584	.574
26.0	0.700	0.685	0.671	0.657	0.644	0.632	0.620	0.608	0.597	0.587	0.576
.1	.702	.688	.673	.660	.647	.634	.622	.610	.600	.589	.578
.2	.705	.690	.676	.662	.649	.637	.625	.613	.602	.591	.580
.3	.708	.693	.679	.665	.652	.639	.627	.615	.604	.593	.583
.4	.710	.696	.681	.667	.654	.642	.629	.617	.606	.596	.585
26.5	.713	.698	.684	.670	.657	.644	.632	.620	.609	.598	.587
.6	.716	.701	.686	.672	.659	.646	.634	.622	.611	.600	.589
.7	.718	.704	.689	.675	.662	.649	.637	.625	.613	.602	.591
.8	.721	.706	.691	.678	.664	.651	.639	.627	.616	.605	.594
.9	.724	.709	.694	.680	.667	.654	.641	.629	.618	.607	.596
27.0	0.727	0.711	0.697	0.683	0.669	0.656	0.644	0.632	0.620	0.609	0.598
.1	.729	.714	.699	.685	.672	.659	.646	.634	.622	.611	.600
.2	.732	.717	.702	.688	.674	.661	.648	.636	.625	.614	.602
.3	.735	.719	.704	.690	.676	.663	.651	.639	.627	.616	.605
.4	.737	.722	.707	.693	.679	.666	.653	.641	.629	.618	.607
27.5	.740	.725	.710	.695	.681	.668	.656	.643	.632	.620	.609
.6	.743	.727	.712	.698	.684	.671	.658	.646	.634	.623	.611
.7	.745	.730	.715	.700	.686	.673	.660	.648	.636	.625	.614
.8	.748	.733	.717	.703	.689	.676	.663	.650	.639	.627	.616
.9	.751	.735	.720	.705	.691	.678	.665	.653	.641	.629	.618
28.0	0.753	0.738	0.722	0.708	0.694	0.680	0.668	0.655	0.643	0.632	0.620
.1	.756	.740	.725	.710	.696	.683	.670	.657	.645	.634	.622
.2	.759	.743	.728	.713	.699	.685	.672	.660	.648	.636	.625
.3	.762	.746	.730	.715	.701	.688	.675	.662	.650	.638	.627
.4	.764	.748	.733	.718	.704	.690	.677	.664	.652	.641	.629
28.5	.767	.751	.735	.720	.706	.693	.679	.667	.655	.643	.631
.6	.770	.754	.738	.723	.709	.695	.682	.669	.657	.645	.633
.7	.772	.756	.740	.726	.711	.697	.684	.671	.659	.647	.636
.8	.775	.759	.743	.728	.714	.700	.687	.674	.662	.650	.638
.9	.778	.762	.746	.731	.716	.702	.689	.676	.664	.652	.640

TABLE 10.—*Concluded.*

THE VALUES OF $\prod_{V}^{V+10\,000} = p_{V} - p_{V+10\,000}$.

p_V	$t_r = 0°$	10°	20°	30°	40°	50°	60°	70°	80°	90°	100°
29.0	0.780	0.764	0.748	0.733	0.719	0.705	0.691	0.678	0.666	0.654	0.642
.1	.783	.767	.751	.736	.721	.707	.694	.681	.668	.656	.645
.2	.786	.769	.753	.738	.724	.710	.696	.683	.671	.659	.647
.3	.788	.772	.756	.741	.726	.712	.699	.685	.673	.661	.649
.4	.791	.775	.759	.743	.729	.714	.701	.688	.675	.663	.651
29.5	.794	.777	.761	.746	.731	.717	.703	.690	.678	.666	.653
.6	.797	.780	.764	.748	.733	.719	.706	.692	.680	.668	.656
.7	.799	.783	.766	.751	.736	.722	.708	.695	.682	.670	.658
.8	.802	.785	.769	.753	.738	.724	.710	.697	.685	.672	.660
.9	.805	.788	.771	.756	.741	.727	.713	.699	.687	.675	.662
30.0	0.807	0.791	0.774	0.758	0.743	0.729	0.715	0.702	0.689	0.677	0.665
.1	.810	.793	.777	.761	.746	.731	.718	.704	.691	.679	.667
.2	.813	.796	.779	.763	.748	.734	.720	.706	.694	.681	.669
.3	.815	.798	.782	.766	.751	.736	.722	.709	.696	.684	.671
.4	.818	.801	.784	.769	.753	.739	.725	.711	.698	.686	.673
30.5	.821	.804	.787	.771	.756	.741	.727	.713	.701	.688	.676
.6	.823	.806	.789	.774	.758	.744	.730	.716	.703	.690	.678
.7	.826	.809	.792	.776	.761	.746	.732	.718	.705	.693	.680
.8	.829	.812	.795	.779	.763	.748	.734	.720	.707	.695	.682
.9	.832	.814	.797	.781	.766	.751	.737	.723	.710	.697	.684

TABLE 11.

THE VALUES OF $\prod_{18\,550}^{20\,000} = p_{18\,550} - p_{20\,000}$.

$p_{18\,550}$	$t_r = 0°$	10°	20°	30°	40°	50°	60°	70°	80°	90°	100°
	Inch.	Inch.	Inch.	Inch.	Inch.	Inch.	Inch.	Inch.	Inch.	Inch.	Inch.
24.0	0.095	0.093	0.091	0.089	0.087	0.085	0.084	0.082	0.081	0.079	0.078
25.0	.099	.097	.095	.093	.091	.089	.087	.086	.084	.083	.081
26.0	.103	.100	.098	.096	.094	.093	.091	.089	.087	.086	.084
27.0	.107	.104	.102	.100	.098	.096	.094	.093	.091	.089	.087
28.0	.111	.108	.106	.104	.102	.100	.098	.096	.094	.092	.091
29.0	.115	.112	.110	.108	.105	.103	.101	.099	.097	.096	.094
30.0	.119	.116	.113	.111	.109	.107	.105·	.103	.101	.099	.097
31.0	.122	.120	.117	.115	.113	.110	.108	.106	.104	.102	.100

As an illustration of the way in which Tables 9, 10 and 11 are to be used let it be supposed that the following values of t_r have been deduced from balloon observations made during static atmospheric conditions :

Between $V = 18\,550$ and $V = 20\,000$, $t_r = 67.0$

" $V = 20\,000$ " $V = 30\,000$, $t_r = 69.5$

" $V = 30\,000$ " $V = 40\,000$, $t_r = 73.0$

" $V = 40\,000$ " $V = 50\,000$, $t_r = 74.0$

" $V = 50\,000$ " $V = 60\,000$, $t_r = 73.5$

Between $V = $ 60 000 and $V = $ 70 000, $t_r = 71.5$
" $V = $ 70 000 " $V = $ 80 000, $t_r = 70.5$
" $V = $ 80 000 " $V = $ 90 000, $t_r = 69.5$
" $V = $ 90 000 " $V = $ 100 000, $t_r = 68.0$
" $V = $ 100 000 " $V = $ 110 000, $t_r = 65.0$
" $V = $ 110 000 " $V = $ 120 000, $t_r = 62.0$

Assume further that the mercurial barometer at the level surface, $V = 18\,550$, shows a pressure of 28.496 inches.

Table 11 for $p_{18\,550} = 28.496$ and $t_r = 67.0$ gives $p_{18\,550} - p_{20\,000} = 0.098$ inch. Therefore the pressure at the level surface $V = 20\,000$ equals $28.496 - 0.098 = 28\,398$ inch. For $p_{20\,000} = 28.398$ and $t_r = 69.5$ Table 10 gives $\Pi_{20\,000}^{30\,000} = 0.666$, so that $p_{30\,000} = 28.398 - 0.666 = 27.732$ inches. Again when $p_{30\,000} = 27.732$ and $t_r = 73.0$ Table 10 gives $\Pi_{30\,000}^{40\,000} = 0.645$, whence $p_{40\,000} = 27.732 - 0.645 = 27.087$. Proceeding upward in this same manner, the following values of $\Pi_{V_0}^{V_1}$ and p_V are obtained:

$\Pi_{20\,000}^{30\,000} = 0.666$ inch	$p_{20\,000} = 28.398$ inches
$\Pi_{30\,000}^{40\,000} = 0.645$ "	$p_{30\,000} = 27.732$ "
$\Pi_{40\,000}^{50\,000} = 0.629$. "	$p_{40\,000} = 27.087$ "
$\Pi_{50\,000}^{60\,000} = 0.614$ "	$p_{50\,000} = 26.458$ "
$\Pi_{60\,000}^{70\,000} = 0.603$ "	$p_{60\,000} = 25.844$ "
$\Pi_{70\,000}^{80\,000} = 0.590$ "	$p_{70\,000} = 25.241$ "
$\Pi_{80\,000}^{90\,000} = 0.577$ "	$p_{80\,000} = 24.651$ "
$\Pi_{90\,000}^{100\,000} = 0.565$ "	$p_{90\,000} = 24.074$ "
$\Pi_{100\,000}^{110\,000} = 0.555$ "	$p_{100\,000} = 23.509$ "
$\Pi_{110\,000}^{120\,000} = 0.545$ "	$p_{110\,000} = 22.954$ "
	$p_{120\,000} = 22.409$ "

From these values of pressure and the corresponding values of t_r already given, may be obtained graphically the mean value of t_r for each pair of the isobaric surfaces $p = 28.5$ in., 28.0 in., 27.5 in., etc., as follows:

Between $p = 28.5$ and $p = 28.0$, $t_r = 68.0$
" $p = 28.0$ " $p = 27.5$, $t_r = 71.0$
" $p = 27.5$ " $p = 27.0$, $t_r = 73.0$
" $p = 27.0$ " $p = 26.5$, $t_r = 74.0$
" $p = 26.5$ " $p = 26.0$, $t_r = 73.5$
" $p = 26.0$ " $p = 25.5$, $t_r = 72.0$
" $p = 25.5$ " $p = 25.0$, $t_r = 71.0$
" $p = 25.0$ " $p = 24.5$, $t_r = 70.0$

Between $p = 24.5$ and $p = 24.0$, $t_r = 69.5$
" $p = 24.0$ " $p = 23.5$, $t_r = 67.5$
" $p = 23.5$ " $p = 23.0$, $t_r = 65.0$
" $p = 23.0$ " $p = 22.5$, $t_r = 62.5$

For these values of t_r Table 9 gives the following :

$E_{28.5}^{28.0} = 7\ 450$ $E_{26.5}^{26.0} = 8\ 100$ $E_{24.5}^{24.0} = 8\ 700$

$E_{28.0}^{27.5} = 7\ 620$ $E_{26.0}^{25.5} = 8\ 230$ $E_{24.0}^{23.5} = 8\ 850$

$E_{27.5}^{27.0} = 7\ 800$ $E_{25.5}^{25.0} = 8\ 380$ $E_{23.5}^{23.0} = 9\ 000$

$E_{27.0}^{26.5} = 7\ 960$ $E_{25.0}^{24.5} = 8\ 530$ $E_{23.0}^{22.5} = 9\ 150$

Finally, to calculate the quantities $V_{31.0}$, $V_{30.5}$, $V_{30.0}$, $V_{29.5}$, etc., we must first determine the number V_{p_1} of level surfaces of gravity lying between sealevel and the first of the isobaric surfaces just named which the balloon meets as it rises into the air. This number consists of two parts, viz., V_0 = the number of level surfaces lying between sealevel and the station-barometer, and $E_{p_0}^{p_1}$ = the number of level surfaces lying between the station-barometer for which the pressure is p_0, and the isobaric surface $p = p_1$. V_0 is a constant and has already been computed for Omaha so that it only remains to obtain the quantity $E_{p_0}^{p_1}$. To accomplish this we use formula (23), written in the following form :

$$E_{p_0}^{p_1} = 1837.3 \times 509.4 \times \log\frac{p_0}{p_1} + 1837.3(t_r - 50° \text{ F.}) \log\frac{p_0}{p_1}.$$

By writing

$$1837.3 \times 509.4 \times \log\frac{p_0}{p_1} = (E_{p_0}^{p_1})_{50}$$

this equation may be written

$$E_{p_0}^{p_1} = (E_{p_0}^{p_1})_{50} + \frac{t_r - 50° \text{ F.}}{509.4}(E_{p_0}^{p_1})_{50}.$$

Table 12 contains the values of $(E_{p_0}^{p_1})_{50}$ considered as a function of p_1 and p_0.

Table 13 contains the values of the expression $\dfrac{t_r - 50}{509.4}(E_{p_0}^{p_1})_{50}$ considered as a function of $(E_{p_0}^{p_1})_{50}$ and t_r. Of course the difference $p_0 - p_1$ never exceeds 0.5 inch.

In the illustrative example for Omaha, $p_0 = 28.496$, $p_1 = 28.0$, and $t_r = 68.0$, whence from Table 12 $(E_{p_0}^{p_1})_{50} = 7\ 130$, and from Table 13, $\dfrac{t_r - 50}{509.4}(E_{p_0}^{p_1})_{50} = +\ 250$. Thus the number of level surfaces lying between the station-barometer and the 28.0-inch isobaric surface equals $7\ 130 + 250 = 7\ 380$. The number V_0 of level surfaces lying between sealevel and the isobaric surface of the station-barometer is 18 550. The total number of level surfaces of gravity included between sealevel and the isobaric surface of 28.0 inches, is therefore $V_{28.0} = 25\ 930$.

If the value $E_{28.0}^{27.5} = 7\ 620$, viz., the number of level surfaces of gravity previously found to lie between the isobaric surfaces $p = 28.0$ and $p = 27.5$, be added to the value 25 930 just found for $V_{28.0}$, then we obtain the quantity $V_{27.5} = 33\ 550$, or .the total number of level surfaces of gravity lying between sealevel and the isobaric surface $p = 27.5$ inches. Again by adding $E_{27.5}^{27.0} = 7\ 800$ to $V_{27.5} = 33\ 550$, we obtain $V_{27.0} = 41\ 350$; by repeating this process the following values of V_p result:

$$V_{28.0} = 25\ 930 \qquad V_{26.0} = 57\ 410 \qquad V_{24.0} = 91\ 250$$
$$V_{27.5} = 33\ 550 \qquad V_{25.5} = 65\ 640 \qquad V_{23.5} = 100\ 100$$
$$V_{27.0} = 41\ 350 \qquad V_{25.0} = 74\ 020 \qquad V_{23.0} = 109\ 100$$
$$V_{26.5} = 49\ 310 \qquad V_{24.5} = 82\ 550 \qquad V_{22.5} = 118\ 250$$

Under static equilibrium in the atmosphere the values of $\Pi_{V_0}^{V}$, p_V, $E_{p_0}^{p}$, and V_p are constants at all points and at all times. Therefore a single balloon ascension, worked up in the manner just described, would suffice to determine for all time the relative positions of the isobaric surfaces and the level surfaces of gravity throughout the whole mass of static atmosphere.

TABLE 12.

THE VALUES OF $\left(E_{p_0}^{p_1}\right)_{30}$, OR THE NUMBER OF LEVEL SURFACES BETWEEN p_0 THE STATION PRESSURE AND p_1 THE PROXIMATE ISOBARIC SURFACE.

$p_1 = 24.5$ Inches.

p_0	0	1	2	3	4	5	6	7	8	9
24.5	0	170	330	500	660	830	990	1160	1320	1490
.6	1660	1820	1990	2150	2320	2480	2650	2810	2980	3140
.7	3300	3470	3630	3800	3960	4130	4290	4450	4620	4780
.8	4950	5110	5270	5440	5600	5770	5930	6090	6260	6420
.9	6580	6750	6910	7070	7230	7400	7560	7720	7890	8050

$p_1 = 25.0$ Inches.

p_0	0	1	2	3	4	5	6	7	8	9
25.0	0	160	320	490	650	810	970	1140	1300	1460
.1	1620	1780	1950	2110	2270	2430	2590	2750	2920	3080
.2	3240	3400	3560	3720	3880	4040	4210	4370	4530	4690
.3	4850	5010	5170	5330	5490	5650	5810	5970	6130	6290
.4	6450	6610	6770	6930	7090	7250	7410	7570	7730	7890

$p_1 = 25.5$ Inches.

p_0	0	1	2	3	4	5	6	7	8	9
25.5	0	160	320	480	640	800	960	1110	1270	1430
.6	1590	1750	1910	2070	2230	2380	2540	2700	2860	3020
.7	3180	3330	3490	3650	3810	3970	4120	4280	4440	4600
.8	4750	4910	5070	5230	5380	5540	5700	5850	6010	6170
.9	6330	6480	6640	6800	6950	7110	7270	7420	7580	7740

$p_1 = 26.0$ Inches.

p_0	0	1	2	3	4	5	6	7	8	9
26.0	0	160	310	470	620	780	940	1090	1250	1400
.1	1560	1720	1870	2030	2180	2340	2490	2650	2800	2960
.2	3110	3270	3420	3580	3730	3890	4040	4200	4350	4510
.3	4660	4820	4970	5130	5280	5440	5590	5740	5900	6050
.4	6210	6360	6510	6670	6820	6970	7130	7280	7440	7690

$p_1 = 26.5$ Inches.

p_0	0	1	2	3	4	5	6	7	8	9
26.5	0	150	310	460	610	770	920	1070	1220	1380
.6	1530	1680	1840	1990	2140	2290	2450	2600	2750	2900
.7	3060	3210	3360	3510	3660	3820	3970	4120	4270	4420
.8	4580	4730	4880	5030	5180	5330	5480	5640	5790	5940
.9	6090	6240	6390	6540	6690	6840	6990	7150	7300	7450

$p_1 = 27.0$ Inches.

p_0	0	1	2	3	4	5	6	7	8	9
27.0	0	150	300	450	600	750	900	1050	1200	1350
.1	1500	1650	1800	1950	2100	2250	2400	2550	2700	2850
.2	3000	3150	3300	3450	3600	3750	3900	4040	·4190	4340
.3	4490	4640	4790	4940	5090	5240	5380	5530	5680	5830
.4	5980	6130	6270	6420	6570	6720	6870	7010	7160	7310

$p_1 = 27.5$ Inches.

p_0	0	1	2	3	4	5	6	7	8	9
27.5	0	150	300	440	590	740	890	1030	1180	1330
.6	1480	1620	1770	1920	2060	2210	2360	2500	2650	2800
.7	2950	3090	3240	3390	3530	3680	3820	3970	4120	4260
.8	4410	4560	4700	4850	4990	5140	5290	5430	5580	5720
.9	5870	6020	6160	6310	6450	6600	6740	6890	7030	7180

TABLE 12.—*Concluded.*

THE VALUES OF $\left(E_{p_0}^{p_1}\right)_{oo}$, OR THE NUMBER OF LEVEL SURFACES BETWEEN p_0 THE STATION PRESSURE, AND p_1 THE PROXIMATE ISOBARIC SURFACE.

$p_1 = 28.0$ *Inches.*

p_0	0	1	2	3	4	5	6	7	8	9
28.0	0	150	290	440	580	730	870	1010	1160	1300
.1	1450	1590	1740	1880	2030	2170	2320	2460	2600	2750
.2	2890	3040	3180	3330	3470	3610	3760	3900	4040	4190
.3	4330	4480	4620	4760	4910	5050	5190	5340	5480	5620
.4	5770	5910	6050	6190	6340	6480	6620	6770	6910	7050

$p_1 = 28.5$ *Inches.*

	0	1	2	3	4	5	6	7	8	9
28.5	0	140	280	430	570	710	850	1000	1140	1280
.6	1420	1570	1710	1850	1990	2130	2280	2420	2560	2700
.7	2840	2980	3130	3270	3410	3550	3690	3830	3970	4110
.8	4260	4400	4540	4680	4820	4960	5100	5240	5380	5520
.9	5660	5810	5950	6090	6230	6370	6510	6650	6790	6930

$p_1 = 29.0$ *Inches.*

	0	1	2	3	4	5	6	7	8	9
29.0	0	140	280	420	560	700	840	980	1120	1260
.1	1400	1540	1680	1820	1960	2100	2240	2380	2510	2650
.2	2790	2930	3070	3210	3350	3490	3630	3770	3910	4040
.3	4180	4320	4460	4600	4740	4880	5010	5150	5290	5430
.4	5570	5710	5840	5980	6120	6260	6400	6530	6670	6810

$p_1 = 29.5$ *Inches.*

	0	1	2	3	4	5	6	7	8	9
29.5	0	140	280	410	550	690	830	960	1100	1240
.6	1380	1510	1650	1790	1920	2060	2200	2340	2470	2610
.7	2750	2880	3020	3160	3290	3430	3570	3700	3840	3980
.8	4110	4250	4390	4520	4660	4790	4930	5070	5200	5340
.9	5470	5610	5750	5880	6020	6150	6290	6420	6560	6700

$p_1 = 30.0$ *Inches.*

	0	1	2	3	4	5	6	7	8	9
30.0	0	140	270	410	540	680	810	950	1080	1220
.1	1350	1490	1620	1760	1890	2030	2160	2300	2430	2570
.2	2700	2840	2970	3100	3240	3370	3510	3640	3780	3910
.3	4040	4180	4310	4450	4580	4710	4850	4980	5120	5250
.4	5380	5520	5650	5780	5920	6050	6190	6320	6450	6590

$p_1 = 30.5$ *Inches.*

	0	1	2	3	4	5	6	7	8	9
30.5	0	130	270	400	530	670	800	930	1060	1200
.6	1330	1460	1600	1730	1860	1990	2130	2260	2390	2520
.7	2660	2790	2920	3050	3190	3320	3450	3580	3710	3850
.8	3980	4110	4240	4370	4510	4640	4770	4900	5030	5160
.9	5300	5430	5560	5690	5820	5950	6080	6220	6350	6480

$p_1 = 31.0$ *Inches.*

	0	1	2	3	4	5	6	7	8	9
31.0	0	130	260	390	520	660	790	920	1050	1180
.1	1310	1440	1570	1700	1830	1960	2090	2220	2350	2480
.2	2610	2740	2870	3000	3130	3260	3390	3520	3650	3780
.3	3910	4040	4170	4300	4430	4560	4690	4820	4950	5080
.4	5210	5340	5470	5600	5730	5860	5990	6120	6250	6370

TABLE 13.

THE VALUES OF $\frac{t_r-50}{509.4}\left(E_{p_o}^{p_1}\right)_{so}$ FOR VALUES OF t_r AND $\left(E_{p_o}^{p_1}\right)_{so}$.

$(E_{p_o}^{p_1})_{so}$	$t_r=0°$	10°	20°	30°	40°	50°	60°	70°	80°	90°	100°
0	0	0	0	0	0	0	0	0	0	0	0
100	— 10	— 10	— 10	0	0	0	0	0	10	10	10
200	— 20	— 20	— 10	— 10	0	0	0	10	10	20	20
300	— 30	— 20	— 20	— 10	— 10	0	10	10	20	20	30
400	— 40	— 30	— 20	— 20	— 10	0	10	20	20	30	40
500	— 50	— 40	— 30	— 20	— 10	0	10	20	30	40	50
600	— 60	— 50	— 40	— 20	— 10	0	10	20	40	50	60
700	— 70	— 50	— 40	— 30	— 10	0	10	30	40	50	70
800	— 80	— 60	— 50	— 30	— 20	0	20	30	50	60	80
900	— 90	— 70	— 50	— 40	— 20	0	20	40	50	70	90
1000	—100	— 80	— 60	— 40	— 20	0	20	40	60	80	100
1100	—110	— 90	— 60	— 40	— 20	0	20	40	60	90	110
1200	—120	— 90	— 70	— 50	— 20	0	20	50	70	90	120
1300	—130	—100	— 80	— 50	— 30	0	30	50	80	100	130
1400	—140	—110	— 80	— 60	— 30	0	30	60	80	110	140
1500	—150	—120	— 90	— 60	— 30	0	30	60	90	120	150
1600	—160	—130	— 90	— 60	— 30	0	30	60	90	130	160
1700	—170	—130	—100	— 70	— 30	0	30	70	100	130	170
1800	—180	—140	—110	— 70	— 40	0	40	70	110	140	180
1900	—190	—150	—110	— 70	— 40	0	40	70	110	150	190
2000	—200	—160	—120	— 80	— 40	0	40	80	120	160	200
2100	—210	—160	—120	— 80	—'40	0	40	80	120	160	210
2200	—220	—170	—130	— 90	— 40	0	40	90	130	170	220
2300	—230	—180	—140	— 90	— 50	0	50	90	140	180	230
2400	—240	—190	—140	— 90	— 50	0	50	90	140	190	240
2500	—250	—200	—150	—100	— 50	0	50	100	150	200	250
2600	—260	—200	—150	—100	— 50	0	50	100	150	200	260
2700	—270	—210	—160	—110	— 50	0	50	110	160	210	270
2800	—270	—220	—160	—110	— 50	0	50	110	160	220	270
2900	—280	—230	—170	—110	— 60	0	60	110	170	230	280
3000	—290	—240	—180	—120	— 60	0	60	120	180	240	290
3100	—300	—240	—180	—120	— 60	0	60	120	180	240	300
3200	—310	—250	—190	—130	— 60	0	60	130	190	250	310
3300	—320	—260	—190	—130	— 60	0	60	130	190	260	320
3400	—330	—270	—200	—130	— 70	0	70	130	200	270	330
3500	—340	—270	—210	—140	— 70	0	70	140	210	270	340
3600	—350	—280	—210	—140	— 70	0	70	140	210	280	350
3700	—360	—290	—220	—150	— 70	0	70	150	220	290	360
3800	—370	—300	—220	—150	— 70	0	70	150	220	300	370
3900	—380	—310	—230	—150	— 80	0	80	150	230	310	380
4000	—390	—310	—240	—160	— 80	0	80	160	240	310	390
4100	—400	—320	—240	—160	— 80	0	80	160	240	320	400
4200	—410	—330	—250	—170	— 80	0	80	170	250	330	410
4300	—420	—340	—250	—170	— 80	0	80	170	250	340	420
4400	—430	—350	—260	—170	— 90	0	90	170	260	350	430
4500	—440	—350	—270	—180	— 90	0	90	180	270	350	440
4600	—450	—360	—270	—180	— 90	0	90	180	270	360	450
4700	—460	—370	—280	—180	— 90	0	90	180	280	370	460
4800	—470	—380	—280	—190	— 90	0	90	190	280	380	470
4900	—480	—380	—290	—190	—100	0	100	190	290	380	480

TABLE 13.—*Concluded.*

The Values of $\dfrac{t_r - 50}{509.4}\left(\mathrm{E}^{p_1}_{p_0}\right)_{so}$ for Values of t_r and $\left(\mathrm{E}^{p_1}_{p_0}\right)_{so}$.

$\left(\mathrm{E}^{p_1}_{p_0}\right)_{so}$	$t_r=0°$	10°	20°	30°	40°	50°	60°	70°	80°	90°	100°
5000	—490	—390	—290	—200	—100	0	100	200	290	390	490
5100	—500	—400	—300	—200	—100	0	100	200	300	400	500
5200	—510	—410	—310	—200	—100	0	100	200	310	410	510
5300	—520	—420	—310	—210	—100	0	100	210	310	420	520
5400	—530	—420	—320	—210	—110	0	110	210	320	420	530
5500	—540	—430	—320	—220	—110	0	110	220	320	430	540
5600	—550	—440	—330	—220	—110	0	110	220	330	440	550
5700	—560	—450	—340	—220	—110	0	110	220	340	450	560
5800	—570	—460	—340	—230	—110	0	110	230	340	460	570
5900	—580	—460	—350	—230	—120	0	120	230	350	460	580
6000	—590	—470	—350	—240	—120	0	120	240	350	470	590
6100	—600	—480	—360	—240	—120	0	120	240	360	480	600
6200	—610	—490	—370	—240	—120	0	120	240	370	490	610
6300	—620	—490	—370	—250	—120	0	120	250	370	490	620
6400	—630	—500	—380	—250	—130	0	130	250	380	500	630
6500	—640	—510	—380	—260	—130	0	130	260	380	510	640
6600	—650	—520	—390	—260	—130	0	130	260	390	520	650
6700	—660	—530	—390	—260	—130	0	130	260	390	530	660
6800	—670	—530	—400	—270	—130	0	130	270	400	530	670
6900	—680	—540	—410	—270	—140	0	140	270	410	540	680
7000	—690	—550	—410	—280	—140	0	140	280	410	550	690
7100	—700	—560	—420	—280	—140	0	140	280	420	560	700
7200	—710	—570	—420	—280	—140	0	140	280	420	570	710
7300	—720	—570	—430	—290	—140	0	140	290	430	570	720
7400	—730	—580	—440	—290	—150	0	150	290	440	580	730
7500	—740	—590	—440	—290	—150	0	150	290	440	590	740
7600	—750	—600	—450	—300	—150	0	150	300	450	600	750
7700	—760	—600	—450	—300	—150	0	150	300	450	600	760
7800	—770	—610	—460	—310	—150	0	150	310	460	610	770
7900	—780	—620	—470	—310	—150	0	150	310	470	620	780

IV. The Relative Positions of the Isobaric Surfaces and the Level Surfaces of Gravity Under Dynamic Conditions.

Experience has shown that the formula for static barometric conditions, viz.,

$$dp = \rho\, dV,$$

also obtains very closely indeed for the actual dynamic conditions. In the succeeding pages I shall assume this formula to hold true since thereby the calculations are simplified and more clearly apprehended.

The primary cause of all atmospheric movements consists in the fact that on account of the unequal heating of the atmosphere the surfaces of equal values of t_r do not coincide with the level surfaces of gravity. The immediate consequence is that the number of isobaric surfaces included between two level surfaces of gravity, as well as the number of the level surfaces included between any pair of isobaric surfaces, can not be everywhere the same, as is the case under static conditions, but on the contrary all the isobaric surfaces are in a state of continuous movement and deformation relative to the level surfaces of gravity, as is well known from the study of daily synoptic weather maps.

Therefore, in order to find the relative positions of the isobaric surfaces and the level surfaces of gravity under dynamic conditions, the quantities $\Pi_{V_0}^V$, p_V, $E_{p_0}^p$, and V_p must be calculated along every vertical in the atmosphere and for every instant. The practical carrying out of this problem would require the sending up simultaneously from a number of stations, kites or balloons carrying self-registers, by means of whose records the four above-mentioned quantities for the verticals at the stations can be calculated. The values thus obtained for these quantities can then be entered on synoptic charts and graphically interpolated, just as is now done, daily, for the barometric readings observed at the meteorological stations and reduced to sealevel.

The kite- and balloon-ascensions heretofore executed may be classed under four types, viz.: ascents reaching great altitudes by means of sounding balloons, as at Trappes, near Paris; ascents in manned balloons, such as are made in Germany; ascents to great heights by means of kites, as at Blue Hill, Mass., and Trappes; and finally the kite-ascents carried out by the Weather Bureau from a large number of specially equipped kite-stations, e. g., the 17 kite-stations of 1898. In coöperation with the manned balloon ascents in Germany, frequent simultaneous ascents of manned and unmanned balloons are carried out at many other European stations (i. e., the international balloon-ascensions). These international balloon-ascensions in Europe and the kite-ascensions made by the U. S. Weather Bureau in America, are especially adapted to synoptic presentation of the four quantities $E_{p_0}^p$, $\Pi_{V_0}^V$, p_V and V_p in the atmosphere, because the pressure may be calculated from them along a number of verticals in the atmosphere for the same moment of time. In the present paper I shall work up only the observations with kites executed by the U. S. Weather Bureau.

For the purpose of synoptical study of the Weather Bureau kite-observations it is very desirable that they be carried out at those hours for which the daily weather maps are made, viz., at 8 A. M. and at 8 P. M., 75th meridian time. Since, however, the wind-conditions often made it impracticable to send up the kite at so early or so late an hour, therefore the observations made at any time during the day must be extrapolated to 8 A. M. or to 8 P. M. The rules for this extrapolation can be deduced only after the proper study of all the kite-observations heretofore made.

Because of our ignorance of these rules I have in the succeeding calculations interpolated to 8 A. M. only those observations obtained from ascents between 6 A. M. and 11 A. M.

The extrapolation of the observations to 8 A. M. or to 8 P. M. and the calculation of the values of the four quantities $\Pi_{V_o}^{V_r}$, p_V, $E_{p_o}^{p_r}$, V_p, can be most advantageously performed by the kite-observers immediately upon reeling in the kite. The results may be readily concentrated to two or three numbers and thus easily telegraphed to the Central Office. As an illustrative example I proceed to show how the kite-ascension at Omaha, Nebr., on 23 Sept., 1898, should be worked up. In Table 14 the figures for pressure (p), temperature (t), and relative humidity (r), are taken from the corresponding curves of the self-recording meteorograph at the kite, while the heights (h) are calculated trigonometrically from the length of the kite-line of steel wire and the angular elevation of the kite. The values of t_r are deduced from p, t and r; and the values of V from the observed elevations, in the manner already described.

TABLE 14.

KITE OBSERVATIONS WITH THE VALUES OF t_r AND V, AT OMAHA, SEPT. 23, 1898.

Time.*	p	t	r	h	t_r	V
	Inch.	° F.	Per cent.	Feet.	° F.	
7⁵⁰ a. m.	28.50	63.0	88	0	66.5	18550
8⁰⁶	27.35	69.5	82	1467	74.0	40490
8¹⁹	27.10	70.0	79	1742	74.5	• 44590
11²⁵	24.80	68.0	51	4453	71.0	85110
11⁴⁵	24.20	68.0	30	5111	69.5	94940
11⁵⁴	23.75	65.0	18	5739	66.0	104340
12¹⁸ p. m.	23.40	64.0	12	6224	64.5	111580
12²⁵	23.15	62.0	11	6541	62.5	116310
12⁴⁷	23.00	61.5	10	6780	62.0	119880
12⁵⁷	22.90	61.0	10	6905	61.5	121750
1⁴⁴	24.10	70.0	5	5131	70.5	95240
1⁵⁷	24.25	71.0	4	4960	71.5	92690
3⁵⁶	25.10	69.0	50	3736	72.0	74400
4¹⁶	25.32	70.0	58	3487	73.5	70680
4²⁵	———	73.0	60	2963	77.0	62840
4³⁰	26.30	77.0	70	2405	82.5	54500
4⁵⁴	26.90	81.0	66	1638	86.0	43040
5³⁵	28.40	87.0	53	0	92.0	18550

Using the values of t_r in Table 14, as abscissæ and the corresponding values of V as ordinates, the points in Fig. 1 are plotted and then a curve drawn through them which gives the values of t_r at the elevation of every level surface of gravity both for the ascent and the descent, by direct reading. By the aid of this (t_r, V)-curve and the observations made at 8 A. M. at the station, the observer or kite official should

* 75th meridian time or 1ʰ 24ᵐ faster than Omaha local mean solar time.

next proceed to construct upon the same set of coördinates by extrapolation, the curve showing the value of t_r at each level surface of the station-vertical, for 8 A. M.   This curve for our example, and as drawn on the same coördinate plane, is shown in Fig. 2,

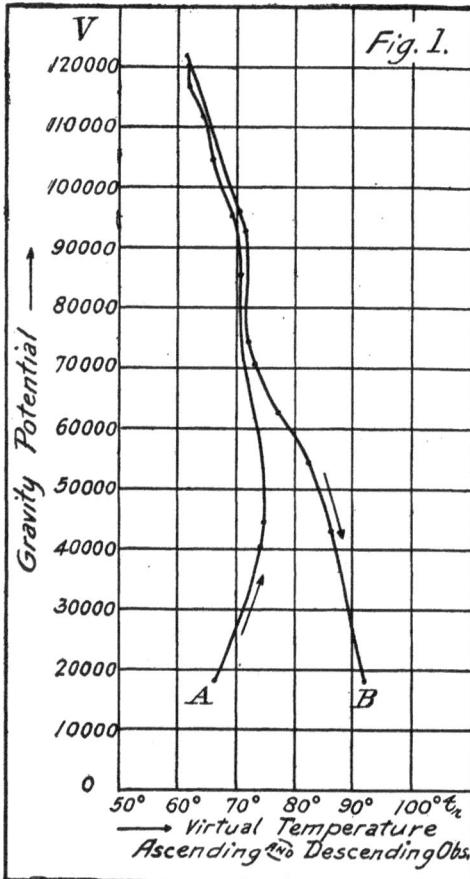

Fig. 1.   The curves of virtual temperatures at Omaha for each value of the gravity potential as calculated from kite records for September 23, 1898. Ascending curve A, descending curve B.

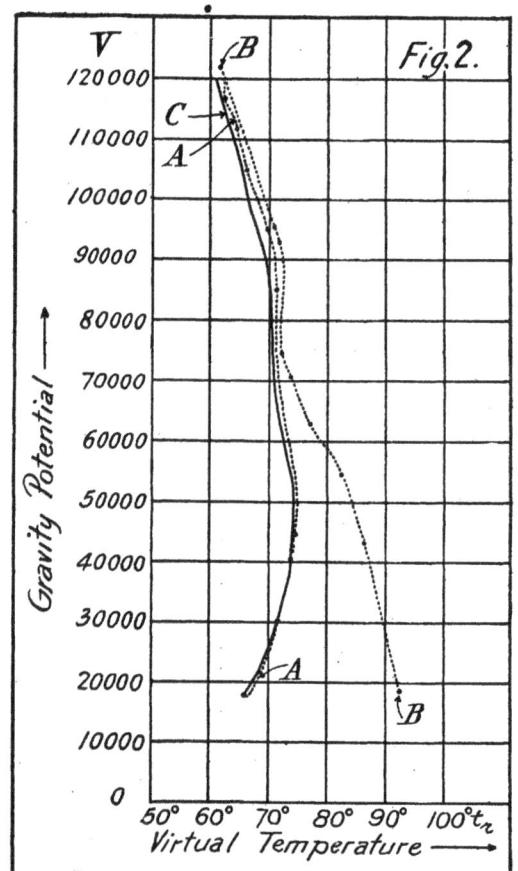

Fig. 2.   The curves of virtual temperatures at Omaha from Fig. 1 with the interpolated curve C for the hour of the synoptic map, or 8 a. m., 75th meridian time, September 23, 1898.

where the 8 A. M. extrapolated (t_r, V)-curve is given as the heavy line (C) together with the curves in dotted lines, obtained directly from the observations of the day as already shown in Fig. 1.   From the extrapolated (t_r, V)-curve of Fig. 2 for 8 A. M. may now be read off the following values for the average virtual temperatures (t_r) at 8 A. M. of the day in question.

Between $V = 18\,550$ and $V = 20\,000$, $t_r = 67°.0$
" $V = 20\,000$ " $V = 30\,000$, $t_r = 69°.5$
" $V = 30\,000$ " $V = 40\,000$, $t_r = 73°.0$
" $V = 40\,000$ " $V = 50\,000$, $t_r = 74°.0$
" $V = 50\,000$ " $V = 60\,000$, $t_r = 73°.5$
" $V = 60\,000$ " $V = 70\,000$, $t_r = 71°.5$
" $V = 70\,000$ " $V = 80\,000$, $t_r = 70°.5$
" $V = 80\,000$ " $V = 90\,000$, $t_r = 69°.5$
" $V = 90\,000$ " $V = 100\,000$, $t_r = 68°.0$
" $V = 100\,000$ " $V = 110\,000$, $t_r = 65°.0$
" $V = 110\,000$ " $V = 120\,000$, $t_r = 62°.0$

We may further assume that the air pressure shown by the station barometer at 8 A. M. equalled 28.496 inches of mercury.[*]

Now, if the barometric formula for static conditions be assumed as sufficiently exact for the assumed dynamic conditions, then the calculation of the four quantities $\Pi_{V_0}^{V}$, p_V, $E_{p_0}^{p}$ and V_p, will be carried on in exactly the same way for the vertical through Omaha, Nebr., on 23 Sept., 1898, 8 A. M., 75th meridian time, as though the atmosphere had been in a static condition on that day. We might therefore here make use of the tables given in the chapter on static conditions. In order to avoid unnecessary repetition, the values just given for t_r for Omaha, 23 Sept., 1893, 8 A. M., 75th meridian time, have been used as the basis for this illustration of static conditions. The following values were found by the method previously described:

$\Pi_{18\,550}^{20\,000} = 0.098$ $p_{18\,550} = 28.496$

$\Pi_{20\,000}^{30\,000} = 0.666$ $p_{20\,000} = 28.398$

$\Pi_{30\,000}^{40\,000} = 0.645$ $p_{30\,000} = 27.732$

$\Pi_{40\,000}^{50\,000} = 0.629$ $p_{40\,000} = 27.087$

$\Pi_{50\,000}^{60\,000} = 0.614$ $p_{50\,000} = 26.458$

$\Pi_{60\,000}^{70\,000} = 0.603$ $p_{60\,000} = 25.844$

$\Pi_{70\,000}^{80\,000} = 0.590$ $p_{70\,000} = 25.241$

$\Pi_{80\,000}^{90\,000} = 0.577$ $p_{80\,000} = 24.651$

$\Pi_{90\,000}^{100\,000} = 0.565$ $p_{90\,000} = 24.074$

$\Pi_{100\,000}^{110\,000} = 0.555$ $p_{100\,000} = 23.509$

$\Pi_{110\,000}^{120\,000} = 0.545$ $p_{110\,000} = 22.954$

$p_{120\,000} = 22.409$

[*] This station-pressure is to be reduced to standard gravity since this reduction is considered as one of the instrumental corrections, see pp. 33 and 42. The correction to a self-registering aneroid should include this item. — C. A.

Here the quantities $p_{18\,550}$, $p_{20\,000}$, etc., are the barometric pressures at the level surfaces $V = 18\,550$, $V = 20\,000$, etc.   From the $(t_r,\ V)$-curve for 8 A. M. in Fig. 2, we find corresponding values of t_r for the same level surfaces as follows:

For $V_{18\,550}$ $p = 28.496,\ t_r = 67.0$
for $V_{20\,000}$ $p = 28.398,\ t_r = 67.5$
for $V_{30\,000}$ $p = 27.732,\ t_r = 71.0$
for $V_{40\,000}$ $p = 27.087,\ t_r = 74.0$
for $V_{50\,000}$ $p = 26.458,\ t_r = 74.0$
for $V_{60\,000}$ $p = 25.844,\ t_r = 72.5$
for $V_{70\,000}$ $p = 25.241,\ t_r = 71.0$
for $V_{80\,000}$ $p = 24.651,\ t_r = 70.0$
for $V_{90\,000}$ $p = 24.074,\ t_r = 69.0$
for $V_{100\,000}$ $p = 23.509,\ t_r = 66.0$
for $V_{110\,000}$ $p = 22.954,\ t_r = 63.5$
for $V_{120\,000}$ $p = 22.409,\ t_r = 61.0$

By plotting the above values of p and t_r as a system of coördinates in which p is ordinate and the corresponding value of t_r is abscissa, a curve is obtained which shows

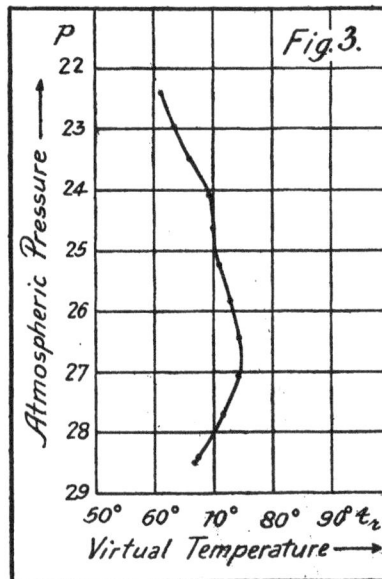

Fig. 3. The $(p,\ t_r)$-curve of virtual temperatures at Omaha for each value of atmospheric pressure as calculated for 8 a. m., 75th meridian time, from the kite record of September 28, 1898.

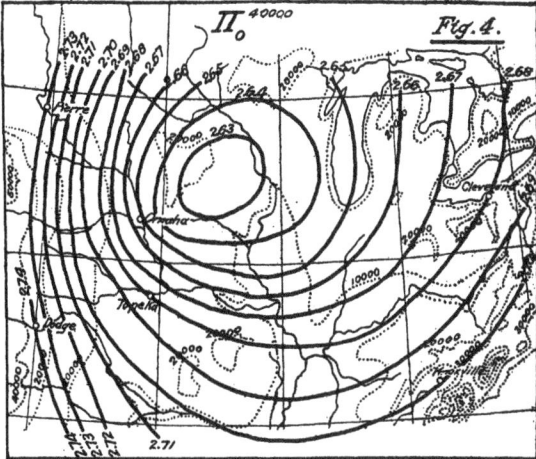

Fig. 4. Chart of $\Pi_0^{40\,000}$ for 8 a. m., September 23, 1898, or lines of equal differences of barometric pressure between sea level and the 40000 potential surface of gravity as telegraphed from all stations to the Central Office.

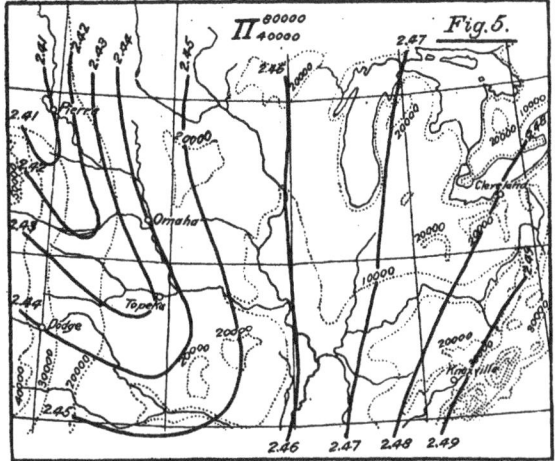

Fig. 5. Chart of $\Pi_{40\,000}^{80\,000}$ for 8 a. m., September 23, 1898, or lines of equal differences of barometric pressure between the 40 000 and 80 000 potential surfaces of gravity as telegraphed to the Central Office.

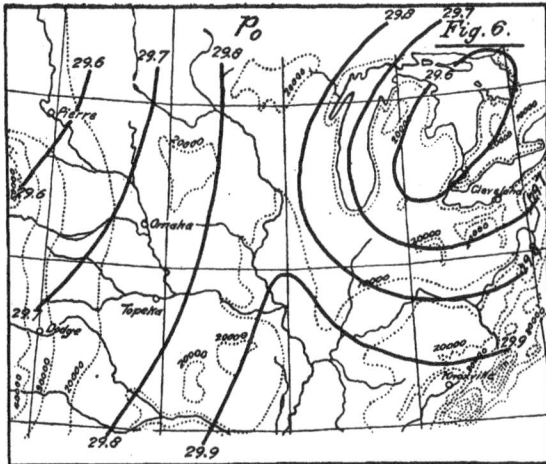

Fig. 6. Chart of p_0 or isobars for sea level for 1898, September 23, 8 a. m., as observed and telegraphed.

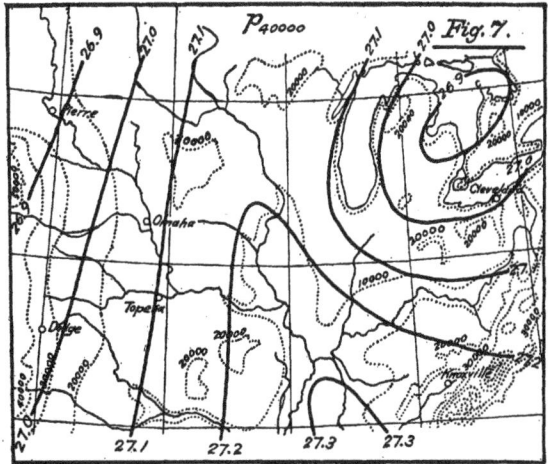

Fig. 7. Chart of $p_{40\,000}$ for 1898, September 23, 8 a. m., or isobars at the 40 000 level surface as deduced from the isobars for sea level by subtracting the numbers on Fig. 4 from those on Fig. 6.

•   the value of t_r in every isobaric surface above Omaha for 23 Sept., 1898, 8 A. M. This curve is shown in Fig. 3.

From this curve the following average values of t_r are easily read off:

Between $p = 28.496$ and $p = 28.000$ $t_r = 68.0$
      " $p = 28.0$ " $p = 27.5$ $t_r = 71.0$
      " $p = 27.5$ " $p = 27.0$ $t_r = 73.0$
      " $p = 27.0$ " $p = 26.5$ $t_r = 74.0$
      " $p = 26.5$ " $p = 26.0$ $t_r = 73.5$
      " $p = 26.0$ " $p = 25.5$ $t_r = 72.0$
      " $p = 25.5$ " $p = 25.0$ $t_r = 71.0$
      " $p = 25.0$ " $p = 24.5$ $t_r = 70.0$
      " $p = 24.5$ " $p = 24.0$ $t_r = 69.5$
      " $p = 24.0$ " $p = 23.5$ $t_r = 67.5$
      " $p = 23.5$ " $p = 23.0$ $t_r = 65.0$
      " $p = 23.0$ " $p = 22.5$ $t_r = 62.5$

For these values of t_r and p we obtain from Table 9

$E_{28.496}^{28.0} = 7\ 380$	$E_{26.5}^{26.0} = 8\ 100$	$E_{24.5}^{24.0} = 8\ 700$
$E_{28.0}^{27.5} = 7\ 620$	$E_{26.0}^{25.5} = 8\ 230$	$E_{24.0}^{23.5} = 8\ 850$
$E_{27.5}^{27.0} = 7\ 800$	$E_{25.5}^{25.0} = 8\ 380$	$E_{23.5}^{23.0} = 9\ 000$
$E_{27.0}^{26.5} = 7\ 960$	$E_{25.0}^{24.5} = 8\ 530$	$E_{23.0}^{22.5} = 9\ 150$

and, by the aid of Tables 12 and 13, the following values

$V_{28.0} = 25\ 930$	$V_{26.0} = 57\ 410$	$V_{24.0} = 91\ 250$
$V_{27.5} = 33\ 550$	$V_{25.5} = 65\ 640$	$V_{23.5} = 100\ 100$
$V_{27.0} = 41\ 350$	$V_{25.0} = 74\ 020$	$V_{23.0} = 109\ 100$
$V_{26.5} = 49\ 310$	$V_{24.5} = 82\ 550$	$V_{22.5} = 118\ 250$

By bringing together the preceding results we may arrange a convenient tabular form as in Table 15 for working up the results of a kite ascension at a kite-station. As an example I have collected in this Table 15, the results already worked out for the observations at Omaha, 23 Sept., 1898.

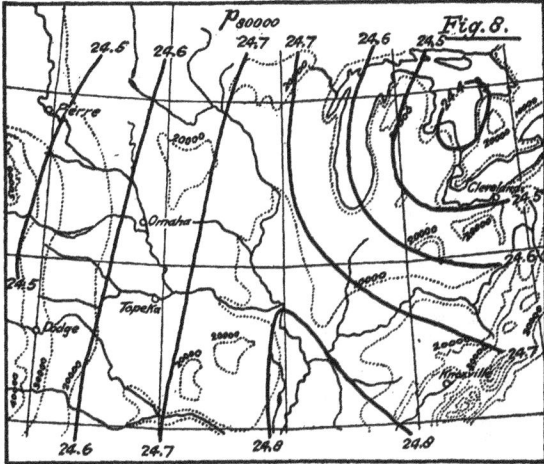

Fig. 8. Chart of $p_{80\,000}$ for 1898, 8 a. m., September 23, or iso-bars at the level surface 80 000 as deduced from the isobars for 40 000 by subtracting the numbers on Fig. 5 from those on Fig. 7.

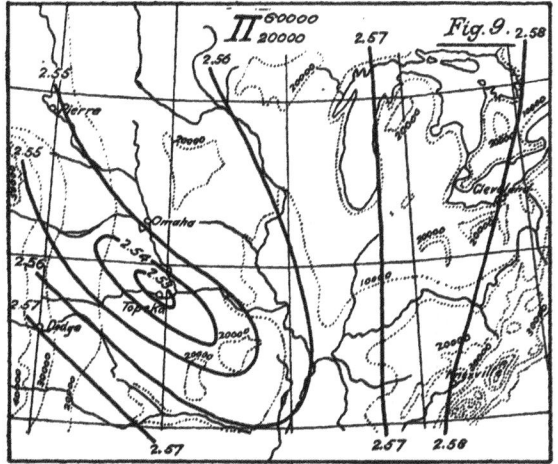

Fig. 9. Chart of $\Pi^{60\,000}_{20\,000}$ for 1898, September 23, 8 a. m., or lines of equal differences of barometric pressure between the 60 000 and the 20 000 potential surfaces of gravity.

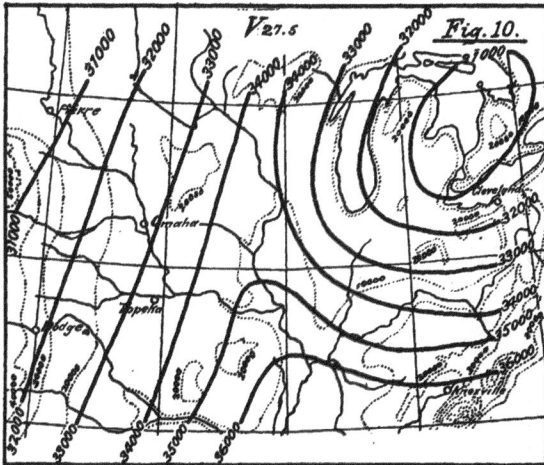

Fig. 10. Chart of $V_{27.5}$ for 8 a. m., September 23, 1898, or chart of the level lines on the isobaric surface 27.5 inches as telegraphed.

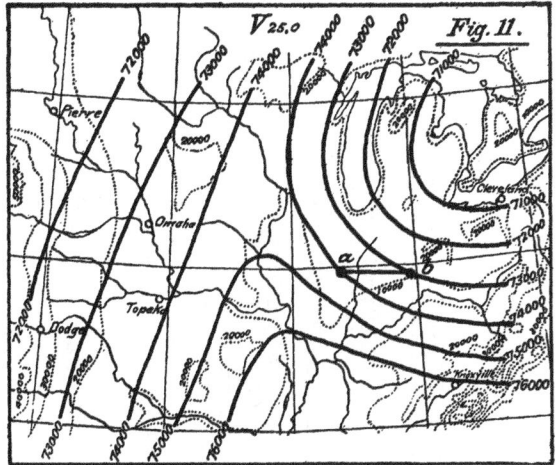

Fig. 11. Chart of $V_{25.0}$ for 8 a. m., September 23, 1898, or chart of the level lines on the isobaric surface 25.0 inches as deduced by adding the numbers on Fig. 12 to those on Fig. 10.

TABLE 15.

FORM FOR THE DYNAMIC COMPUTATIONS BASED ON KITE OBSERVATIONS.
OMAHA, NEBRASKA, SEPT. 23, 1898.

1	Computation of t_r and V.									Computation of $\Pi_{V_0}^{P_1}$ and p_V.					1
2	1	2	3	4	5	6	7	8	9	10	11	12	13	14	2
3	Time.*	Bar.	Temp.	t_1-t	r	t_r-t	t_r	z	V	V	t_r	$\Pi_{V_0}^{P_1}$	p_V	t_r	3
	h m	Inch.	°	°		°	°	feet					Inch.	°	
4	8:00 a. m.	28.496	63.5	4.0	88	3.5	67	0	18 550	18 550	67.0	0.098	28.496	67	4
5	7:50	28.50	63	4.0	88	3.5	66.5	0	18 550	20 000	69.5	0.666	28.398	67.5	5
6	8:06	27.35	69.5	5.5	82	4.5	74	1467	40 490	30 000	73.0	0.645	27.732	71	6
7	8:19	27.10	70	5.5	79	4.5	74.5	1742	44 590	40 000	74.0	0.629	27.087	74	7
8	11:25	24.80	68	5.5	51	3.0	71	4453	85 110	50 000	73.5	0.614	26.458	74	8
9	11:45	24.20	68	5.5	30	1.5	69.5	5111	94 940	60 000	71.5	0.603	25.844	72.5	9
10	11:54	23.75	65	5.0	18	1.0	66	5739	104 340	70 000	70.5	0.590	25.241	71	10
11	12:13 p. m.	23.40	64	5.0	12	0.5	64.5	6224	111 580	80 000	69.5	0.577	24.651	70	11
12	12:25	23.15	62	5.0	11	0.5	62.5	6541	116 310	90 000	68.0	0.565	24.074	69	12
13	12:47	23.00	61.5	4.5	10	0.5	62	6780	119 880	100 000	65.0	0.555	23.509	66	13
14	12:57	22.90	61	4.5	10	0.5	61.5	6905	121 750	110 000	62.0	0.545	22.954	63.5	14
15	1:44	24.10	70	6.0	5	0.5	70.5	5131	95 240	120 000			22.409	61	15
16	1:57	24.25	71	6.5	4	0.5	71.5	4960	92 690						16
17	3:56	25.10	69	5.5	50	3.0	72	3736	74 400						17
18	4:16	25.32	70	6.0	58	3.5	73.5	3487	70 680						18
19	4:25		73	6.5	60	4.0	77	2963	62 840						19
20	4:39	26.30	77	7.5	70	5.5	82.5	2405	54 500						20
21	4:54	26.90	81	8.0	66	5.0	86	1638	43 040						21
22	5:25	28.40	87	9.5	53	5.0	92	0	18 550						22

* All records are kept on 75th meridian time which is 1ʰ 24ᵐ faster than Omaha local mean solar time.

TABLE 15.—*Continued.*

FORM FOR THE DYNAMIC COMPUTATIONS BASED ON KITE OBSERVATIONS.
OMAHA, NEBRASKA, SEPT. 23, 1898.

1	Computation of $E_{P_0}^{P_1}$ and V_p.					Values of t_r in situ.					1
2	15	16	17	18	19	20	21	22	23	24	2
3	p	t_r	$E_{P_0}^{P_1}$	V_p	t_r	V	Time.	t_r	Time.	t_r	3
	Inch.	°					h m	°	h m	°	
4	28.496	68	7 380	18 550	67.0	0	—	—	—	—	4
5	28.0	71	7 620	25 930	69.5	10 000	—	—	—	—	5
6	27.5	73	7 800	33 550	72	20 000	7:52 a. m.	67	5:23 p. m.	91.5	6
7	27.0	74	7 960	41 350	74	30 000	7:57	71	5:10	89.5	7
8	26.5	73.5	8 100	49 310	74	40 000	8:06	74	4:58	87.0	8
9	26.0	72	8 230	57 410	73	50 000	8:45	74.5	4:48	84.0	9
10	25.5	71	8 380	65 640	71.5	60 000	9:34	73.5	4:29	79.0	10
11	25.0	70	8 380	74 020	70	70 000	10:22	71.5	4:12	73.5	11
12	24.5	69.5	8 530	82 550	70	80 000	11:08	71.0	3:23	72.0	12
13	24.0	67.5	8 700	91 250	68.5	90 000	11:36	70.5	2:12	72.0	13
14	23.5	65	8 850	100 100	66	100 000	11:40	67	1:39	69.0	14
15	23.0	62.5	9 000	109 100	63.5	110 000	12:10 p. m.	65	1:30	66.0	15
16	22.5		9 150	118 250	61.5	120 000	12:50	62	1:13	62.0	16
17						130 000	1:00	60	1:00	60.0	17
18											18
19											19
20											20
21											21
22											22

In this schematic presentation, the various columns as numbered have the following significance:

No. 1 contains the moments of observation. All time records are uniformly in 75th meridian time.

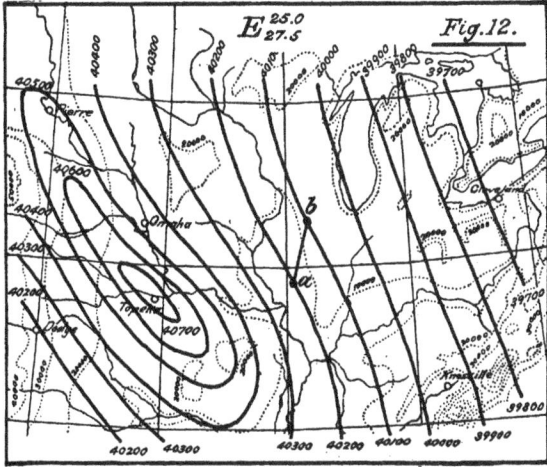

Fig. 12. Chart of $E_{27.5}^{25.0}$ for 8 a. m., September 23, 1898, or the number of solenoids in the layer of atmosphere over any place between the isobaric surfaces 27.5 and 25.0 as computed and telegraphed ; showing the tendency of the air at any place to maintain a vertical circulation.

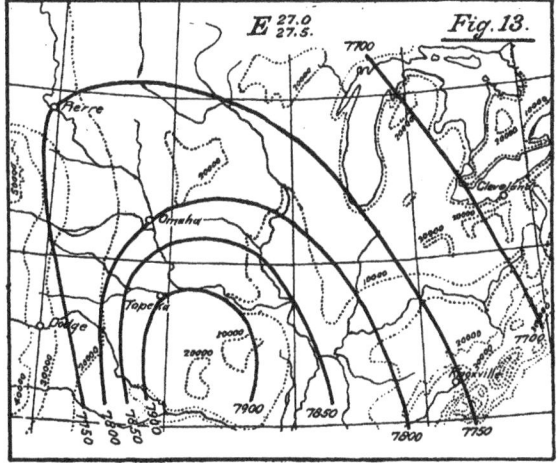

Fig. 13. Chart of $E_{27.5}^{27.0}$ for 8 a. m., September 23, 1898, or the number of solenoids in the layer of atmosphere between the isobars 27.0 and 27.5 above any place.

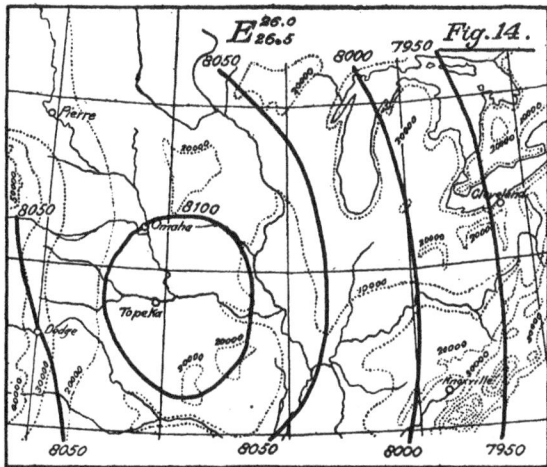

Fig. 14. Chart of $E_{26.5}^{26.0}$ for 8 a. m., September 23, 1898, or the number of solenoids in the layer of atmosphere between the isobars 26.0 and 26.5 above any place.

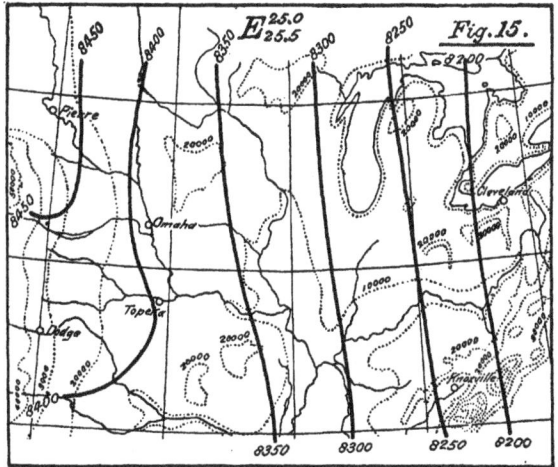

Fig. 15. Chart of the $E_{25.5}^{25.0}$ for 8 a. m., September 23, 1898, or the number of solenoids in the layer of atmosphere between the isobars 25.0 and 25.5 above any place.

Nos. 2 and 3, respectively, the pressure and temperature registered at these hours [the local pressure expressed in inches of mercury under standard gravity.—C. A.].

No. 4, the values of $(t_1 - t)$ for these pressures and temperatures as obtained from Table 7.

No. 5, the registered relative humidities.

No. 6, the values of $(t_r - t)$ deduced from Table 8, for the data in columns 4 and 5.

No. 7, the t_r or the sum of the $(t_r - t)$ in column 6 and the temperatures t in column 3.

No. 8, the observed elevations computed trigonometrically.

No. 9, the values of the gravity potentials V obtained from No. 8 by means of Table 3.

From the t_r and V in columns 7 and 9 the (t_r, V) curve of Fig. 1 is constructed and along side of it the corresponding extrapolated curve for 8 A. M. as in Fig. 2. From the (t_r, V) curve for 8 A. M. we read off the mean values of t_r for the intervals $V = 18\,550$ to $V = 20\,000$, $V = 20\,000$ to $V = 30\,000$, etc., and proceed to the following columns:

No. 10, the ordinal numbers of the level surfaces of gravity.

No. 11, the mean values of t_r for the intervals between the surfaces of column No. 10.

No. 12, the values of $\Pi_{V_0}^{V_1}$ for the average t_r as given by Tables 10 and 11.

No. 13, the value of p_V for each level surface obtained by successive algebraic additions of $\Pi_{V_0}^{V_1}$ to the reading of the station-barometer at 8 A. M.

No. 14, contains the values of t_r for the level surfaces, $V = 18\,550$, $V = 20\,000$, etc. at 8 A. M., obtained directly from the extrapolated (t_r, V)-curve of Fig. 2.

From the values of p_V and t_r given in columns 13 and 14, the curve of Fig. 3 is constructed and from this the mean value of t_r for each half-inch of barometric change is read off.

No. 15, contains the barometric pressure for each of these half-inch intervals.

No. 16 gives the corresponding mean values of t_r.

No. 17 gives the values of $E_{p_0}^{p_1}$ for these t_r-values, obtained by aid of Tables 9, 12 and 13.

No. 18 contains the values of V_p, that result from successive additions of the values in column 17 [to the value of V_p for the level surface that contains the station barometer.—C. A.].

From the curves in Figs. 2 and 3 there may also be determined the values of t_r for the isobaric surfaces at 8 A. M., and for the level surfaces of gravity at the moments when the kite passed through them.

No. 19 contains the values of t_r at 8 A. M. read off from the curve of Fig. 3 and corresponding to the isobaric surfaces given in column 15.

No. 20 gives the ordinal number for each 10 000th level surface of gravity.

Nos. 21 and 23 give the times when the kite passed through each of the surfaces given in column 20, ascending and descending respectively.

These times of passage through the level surfaces as given in columns 21 and 23 may readily be obtained graphically as follows: The times given in column 1 are plotted as abscissæ and the values of V in column 9 as ordinates. Then the (Time, V)-curve is drawn through the points thus plotted and from this curve the time of the moment of intersection for each 10 000th level surface of gravity may be read off directly.

Nos. 22 and 24 give the values of t_r at each passage through the level surfaces of column 20; these values having been read from the curves for the kite ascension shown in Fig. 2.

Preparation of Synoptical Charts at the Central Station.

For synoptic study at the central station it is sufficient to telegraph only some of the most important of the quantities above calculated, e. g., the quantities $\Pi_0^{40\,000}$, $\Pi_{40\,000}^{80\,000}$, $\Pi_{80\,000}^{120\,000}$, $V_{27.5}$, $E_{27.5}^{25.0}$, and $\overline{E}_{25.0}^{22.5}$. The value of $\Pi_0^{40\,000}$ is obtained by subtracting $p_{40\,000} = 27.087$ from the reading of the station-barometer reduced to sealevel, or $p_0 = 29.74$, whence results the difference, $\Pi_0^{40\,000} = 2.653$. In the same way are obtained the values $\Pi_{40\,000}^{80\,000} = 27.087 - 24.651 = 2.436$, and $\Pi_{80\,000}^{120\,000} = 24.651 - 22.409 = 2.242$. The value of $V_{27.5} = 33\,550$ is taken directly from column 18 of Table 15. The values of $E_{27.5}^{25.0} = 40\,470$, and $E_{25.0}^{22.5} = 44\,230$ are the differences $V_{25.0} - V_{27.5}$ and $V_{22.5} - V_{25.0}$ respectively. The numbers to be telegraphed to the central station are therefore 2.653, 2.436, 2.242, 33 550, 40 470 and 44 230. For telegraphic purposes these numbers may be shortened by dropping the first and the last figures of each, so that we have to telegraph only the abbreviated numbers 65, 44, 24, 355, 047 and 423. These may be combined into three groups of five figures each, as for example 65 355, 44 047, 24 423.*

Now assume that all the kite-stations where ascensions were made with registering instruments during the forenoon of 23 Sept., 1898, had worked up their observations according to the foregoing method and sent telegraphic reports to the central office. Then these telegrams as received would have read somewhat as follows:

23 SEPT., 1898, 8 A. M., 75TH MERIDIAN TIME.

Cleveland, O.	68 135	48 963	....
Dodge City, Kan.	74 193	44 016	22 446
Knoxville, Tenn.	70 635	49 993	....
Omaha, Nebr.	65 355	44 047	24 423
Pierre, S. D.	73 076	41 054	....
Topeka, Kan.	68 363	43 074	....

* This contraction for economy in European telegraphy would be advantageously replaced in America by our usage of short cipher code words or syllables. — C. A.

At the central office of the Weather Bureau by means of these numbers charts can be drawn presenting synoptically the values of $\Pi_0^{40\,000}$, $\Pi_{40\,000}^{80\,000}$, $p_{40\,000}$, $p_{80\,000}$, $E_{27.5}^{25.0}$, $V_{27.5}$ and $V_{25.0}$. The first step is to separate the figures of the telegrams and to supply the missing figures, with the following result:

23 SEPT., 1898. 8 A. M.

Obs. Station.	$\Pi_0^{40\,000}$	$\Pi_{40\,000}^{80\,000}$	$\Pi_{80\,000}^{120\,000}$	$V_{27.5}$	$E_{27.5}^{25.0}$	$E_{25.0}^{22.5}$
Cleveland, O.	2.68	2.48	—	31 350	39 630	—
Dodge City, Kans.	2.74	2.44	2.22	31 930	40 160	44 460
Knoxville, Tenn.	2.70	2.49	—	36 350	39 930	—
Omaha, Nebr.	2.65	2.44	2.24	33 550	40 470	44 230
Pierre, So. Dak.	2.73	2.41	—	30 760	40 540	—
Topeka, Kans.	2.68	2.43	—	33 630	40 740	—

The second step is to enter these values at the appropriate stations on a series of skeleton maps. The sketch map forming Fig. 4 on page 67 gives a synoptic map of the quantity $\Pi_0^{40\,000}$. Fig. 5, page 67, shows a similar map for the quantity $\Pi_{40\,000}^{80\,000}$. The maps, Figs. 4 and 5 and curves have been drawn just as the isobars for sealevel are drawn on the usual isobaric maps. The three maps following, viz., Figs. 6, 7, 8, pages 67, 69, show the quantities p_0, $p_{40\,000}$, $p_{80\,000}$, respectively. The map for p_0, Fig. 6, is copied directly from the Weather Bureau map of barometric pressure reduced to sealevel. The $p_{40\,000}$ map, Fig. 7, which is a map of the isobars at the level surface $V = 40\,000$, is constructed graphically by superposition of the p_0 chart, Fig. 6, and the $\Pi_0^{40\,000}$ chart, Fig. 4, making use of the relation

$$p_{40\,000} = p_0 - \Pi_0^{40\,000}.$$

The $p_{80\,000}$ map, Fig. 8, page 69, is constructed in an analogous way by superposing Figs. 5 and 7, using the relation

$$p_{80\,000} = p_{40\,000} - \Pi_{40\,000}^{80\,000}.$$

The synoptic map of the values $\Pi_{20\,000}^{60\,000}$ forming Fig. 9, of page 69, will be discussed later.

Fig. 10, on page 71, shows the synoptic distribution of the quantity $V_{27.5}$, $i.\ e.$, the number of level surfaces of gravity between sealevel and the isobaric surface for $p = 27.5$ inches; it is constructed from the telegraphed values of $V_{27.5}$ superposed on the map of isobars for sealevel. The last map on page 69, Fig. 11, shows the distribution of the values of $V_{25.0}$, $i.\ e.$, the number of level surfaces between sealevel and the isobaric surface $p = 25.0$. It is constructed by superposing Fig. 10 for $V_{27.5}$ and Fig. 12 for $E_{27.5}^{25.0}$ using the relation

$$V_{25.0} = V_{27.5} + E_{27.5}^{25.0}.$$

The first map on page 71, viz., Fig. 12, presents a synoptic view of the values of the quantity $E^{25.0}_{27.5}$, and is constructed from the telegraphed values of $E^{25.0}_{27.5}$ in a manner analogous to the chart of $\Pi^{40\,000}_0$, page 67, Fig. 4. The remaining maps on page 71, viz., Figs. 13, 14, 15, present synoptic views of the distribution of the quantities $E^{27.0}_{27.5}$, $E^{26.0}_{26.5}$, and $E^{25.0}_{25.5}$, respectively, and will be discussed later.

The distribution of pressure under the prevailing dynamic conditions in the atmosphere is thus presented on the one hand by p_V charts, showing the isobars on the level surfaces of gravity, and on the other hand by V_p charts, showing the level lines of gravity on the isobaric surfaces. These two systems of charts taken together present a very clear picture of the relative positions of the isobaric surfaces and of the level surfaces of gravity. From kite observations and by the aid of the tables accompanying this memoir, isobars on the level surfaces of gravity can be constructed for much smaller intervals, i. e., for the level surfaces of $V = 0$, $V = 10\,000$, $V = 20\,000$, $\cdots V = 180\,000$, as also level lines on the isobaric surfaces of $p = 31$, $p = 30.5$, $p = 30.0 \cdots p = 19.0$. The charts on pages 67, 69, 71, however, suggest that such intervals are much too small. In fact, the charts for $p_{80\,000}$, $p_{40\,000}$, and p_0 show nearly the same characteristics; and the same is true of the charts for $V_{25.0}$ and $V_{27.5}$. It is obviously superfluous to draw charts for such small intervals that the types are nearly identical. On the other hand the interval must not be too large since then the features would differ so much that it would be difficult or impossible to follow the continuity of the change in the type with increasing elevation. We must learn through experience what intervals are to be chosen as best suited to our studies, and to the condition of the atmosphere.

I have chosen the isobaric map drawn for sealevel as the base for the p_V- and V_p-maps, because the values of atmospheric pressures as telegraphed from permanent observing stations are, without exception, reduced to sealevel. But when one wishes to construct maps for the free atmosphere, it is quite superfluous to first reduce the pressure to sealevel, and then re-reduce it upwards from sealevel to a higher one. The rational way would be to reduce the pressures observed at the permanent stations, not to sealevel but to the nearest level surface of gravity for which a p_V-map is to be constructed, and then use the value of p_V thus obtained in constructing the corresponding p_V-map. In an analogous way the number of level surfaces of gravity lying between the level of the station-barometer and the nearest isobaric surface adopted for mapping values of V_p, might be calculated; whence by adding the values of V_0, the values of V_p for the isobaric surface in question could be determined and be used in constructing the proper V_p-map. The values of p_V and V_p obtained from the kite-observations would thus serve in constructing their respective maps for the free air and the values of $\Pi^{V_1}_{V_0}$ and $E^{p_1}_{p_0}$ could be used in the manner already described, for superposition on the p_V- and V_p-maps. By the foregoing method of procedure, however, no isobaric charts at sealevel would be obtained for those regions where the stations are at considerable altitudes above sealevel.

V. The Dynamic Significance of the Charts of p_V, V_p; $E_{p_o}^{p}$ and $\Pi_{V_o}^{V}$.

The following conclusions are deduced on the distinct assumption that the earth does not rotate and that friction does not exist. I defer to a later paper the consideration of the influence of the rotation of the earth and of friction upon the dynamic processes of the atmosphere. In this section I shall consider only the primary cause of all atmospheric movements, in other words the want of uniformity as to temperature and humidity. This is that which has the power to set up a movement in an atmosphere otherwise at rest relative to the earth, whereas the earth's rotation and the friction do not possess such power.

Significance of p_V-maps.—The dynamic significance of the p_V-maps, namely, the maps of the isobars on the different level surfaces of gravity, is already familiar enough through the daily use of the maps of the isobars at sealevel. I would only here call attention to the fact that in order to obtain the acceleration of the particles of air the pressure-gradient must be divided by the appropriate density of the air. Consequently, in the higher levels where the air has a less density, the same gradient of pressure will produce a much greater velocity than it would at sealevel.

Significance of V_p-maps. — The dynamic significance of the V_p-maps (which may be called topographic charts of isobaric surfaces, or maps showing the intersections of an isobaric surface by successive level surfaces of equal values of gravity), is seen from the fact that an air-particle moving on such an isobaric surface experiences the same acceleration as if it were confined to that surface and subject only to the force of gravity. Therefore, if we assume that an air-particle moves from a to b on the $V_{25.0}$-chart (see Fig. 11, page 69), and during this movement remains in the isobaric surface, $p = 25.0$, then the acceleration of the particle may be found by dividing the difference in gravity-potential at the points a and b by the length of the path of the particle or the distance between a and b. Now the gravity-potential at a equals $V_a = 74\,000\,\dfrac{\text{mile}^2}{\text{hour}^2}$, and at b equals $V_b = 73\,000\,\dfrac{\text{mile}^2}{\text{hour}^2}$, so that the difference in gravity-potential at the two points is $V_a - V_b = 1\,000\,\dfrac{\text{mile}^2}{\text{hour}^2}$. The distance between a and b is approximately 140 miles, whence the acceleration of the particle of air is seen to be $\dfrac{1\,000}{140} = 7.14\,\dfrac{\text{mile}}{\text{hour}^2}$. It is easy to calculate the velocity v_1 of the air-particle, when it arrives at b, from the velocity v_0 it had at a and the difference in gravity-potential, $V_a - V_b$, by the aid of the well known formula

$$\frac{v_1^2 - v_0^2}{2} = V_a - V_b.$$

Thus if it be assumed that the velocity v_0 at the point a be $10 \frac{\text{mile}}{\text{hour}}$ and that $V_a - V_b = 1\,000 \frac{\text{mile}^2}{\text{hour}^2}$ then the velocity v_1 of the particle on arriving at b is obtained by solving the equation

$$v_1^2 = 10^2 + 2 \times 1\,000 = 2\,100$$

$$v_1 = \sqrt{2\,100} = 45.8 \frac{\text{mile}}{\text{hour}}.$$

This method of using the map for calculating the acceleration of an air-particle from the length of its path and the difference in gravity-potential, and for calculating the velocity of the particle from the difference in gravity-potential and the initial velocity, may also be used when we consider relative movements, since the component of acceleration due to the Earth's rotation always acts in a direction at right angles to the path of the particle and thus has no effect upon the acceleration along this path.

The calculations have been carried out for a particle which always remains in the same isobaric surface. They are, however, equally applicable to particles moving within a slight distance from the given isobaric surface, because these surfaces, which lie very close to one another, have almost mutually parallel directions, and thus intersect very nearly the same number of level surfaces of gravity.

Comparison of V_p- and p_γ-maps.—It seems to me that from a dynamic point of view the V_p-maps possess certain advantages over the p_γ-maps. These advantages arise, partly, from the fact that the acceleration and the square of the velocity of a particle may be read directly from the V_p-maps without taking into consideration the density of the air, whereas the pressure-gradients obtained from the p_γ-maps must first be divided by the density of the air in order to obtain these quantities. When we limit ourselves to purely qualitative considerations these advantages appear yet more striking; for the accelerations are directly proportional to the number of lines [between any two points] on the V_p charts and quite independent of altitude in the atmosphere. On the other hand, if the p_γ-maps for two different levels show the same number of lines [within the same distance], then the air-particles at the higher level have the greater acceleration. It is thus seen that the V_p-maps for different levels are completely comparable with one another, while the p_γ-maps are not.

Significance of $E_{p_0}^{p_1}$-*maps.* — The dynamic significance of the $E_{p_0}^{p_1}$-maps, Figs. 12–15, results from a principle in hydrodynamics recently stated by Prof. V. Bjerknes,* and I would first recall this principle. According to Lord Kelvin's definition, the circulation of a closed curve made up of atmospheric particles, consists of the sum of the tangential components of the velocity of every particle around the whole curve. If the velocity of a particle of the curve be designated by u, and the tangential component of this velocity along the curve by u_t, then the circulation "C" is expressed by the integral

$$C = \int u_t \delta s$$

where "δs" is a longitudinal element of the curve and the integration is to be carried out completely around the whole of the closed curve. This "circulation" is an expression for the rotatory movement of the atmosphere, for wherever the velocity of the air has a potential, there all closed curves have no "circulation"; and conversely, the more intense is the rotatory movement of the air so much the greater is the "circulation" of the closed curves.

By means of the integral just cited, the "circulation" of a closed curve in the atmosphere may be determined from simultaneous observations of the direction and velocity of the wind at different points on the curve. Bjerknes has given a theorem for calculating the increase or decrease of the "circulation" during a unit of time, by using the observations of pressure, temperature and humidity at points along the curve. If then we have the four elements — wind, pressure, temperature and relative humidity observed at any moment of time, for various points along a closed curve in the atmosphere we may calculate the "circulation" of that curve not only for the moment of observation, but also for a series of instants both preceding and following that moment. The theorem may be mathematically formulated as follows:

$$\frac{dC}{dt} = - \int v \, dp = A. \qquad (25)$$

Here dC/dt is the increase of circulation C in a unit of time; v is the specific volume of a particle of air on the curve, and p is the pressure prevailing at this particle. The integration is to be carried out around the whole closed curve and will give $A =$ the number of solenoids,† enclosed within the closed curve. The law may then be stated as follows.

*See V. Bjerknes. "The dynamic principle of circulatory movements in the atmosphere."—Monthly Weather Review, Oct., 1900, p. 434.

† A solenoid is a tubular figure in the atmosphere arising from the intersections of surfaces of equal pressure, or isobaric surfaces, with surfaces of equal specific volume, or isosteric surfaces. The unit solenoid is found between two isobaric surfaces differing by the unit of pressure and two isosteric surfaces differing by the unit of specific volume.

The increase of circulation per unit of time, in a closed atmospheric curve made up of air-particles is equal to the total number of unit solenoids embraced within that curve.

Now the number and position of the solenoids in the atmosphere may be obtained in a very simple way from the $E_{p_2}^{p_1}$ maps. Thus we choose any two points a and b on any two of the lines of such a map as the $E_{27.5}^{25.0}$ map shown in Fig. 12, page 71. Imagine verticals falling from these points in the atmosphere to corresponding points on the isobaric surfaces $p = 27.5$ and 25.0 which vertical lines we will designate also by the letters a and b. The lower ends of these verticals are connected by the line a–b, which lies wholly in the isobaric surface 25.0 and the upper ends are connected by the line a–b which lies wholly in the isobaric surface $p = 27.5$. Thus is obtained a closed curve in the atmosphere consisting of two vertical portions aa and bb, and two isobaric portions, ab and ba. The number of solenoids within this closed curve may be determined by carrying out the integration $\int v dp$ around the whole curve. Now along the two isobaric portions ab and ba of the curve, both vdp and $\int v dp$, are equal to zero so it only becomes necessary to perform the integration along the two verticals aa and bb. The integral along aa may be represented by $\left(\int_{25.0}^{.5} v \cdot dp\right)_a$ and the integral along bb by $\left(\int_{25.0}^{27.5} v \cdot dp\right)_b$, then by virtue of equation (25) we have

$$A = \left(\int_{25.0}^{27.5} v \cdot dp\right)_a - \left(\int_{25.0}^{27.5} v \cdot dp\right)_b \qquad (26)$$

which integral may be simplified by making use of the barometric formula *

$$dV = - v \cdot dp.$$

By integrating both sides of this latter formula along the vertical aa we find that

$$V_{25.0} - V_{27.5} = \left(\int_{25.0}^{27.5} v \cdot dp\right)_a.$$

If by $(E_{27.5}^{25.0})_a$ we designate the number of level surfaces of gravity lying between the 27.5- and 25.0-isobaric surfaces along the vertical a, then we may write

$$\left(\int_{25.0}^{27.5} v \cdot dp\right)_a = (E_{27.5}^{25.0})_a.$$

Whence from (7) we have

$$(V_{25.0} - V_{27.5})_a = (E_{27.5}^{25.0})_a.$$

Analogously we find that

$$\left(\int_{25.0}^{27.5} v \cdot dp\right)_b = (E_{27.5}^{25.0})_b.$$

* See equations (1) and (10).

By substituting these into (26) there results

$$A = (E_{27.5}^{25.0})_a - (E_{27.5}^{25.0})_b \tag{27}$$

This formula holds true for any two points a and b on the $E_{27.5}^{25.0}$-map and for the corresponding closed curves in the atmosphere. For the points a and b shown on the $E_{27.5}^{25.0}$-map (Fig. 12) of page 71 we have $(E_{27.5}^{25.0})_a = 40\ 200\ \dfrac{mile^2}{hour^2}$ and $(E_{27.5}^{25.0})_b = 40\ 100\ \dfrac{mile^2}{hour^2}$, so that by equation 27, $A = 100\ \dfrac{mile^2}{hour^2}$. If now we move the points a and b of this map at will along the curves 40 200 and 40 100 respectively, and imagine the closed curve consisting of the verticals a and b, and the connecting lines lying in the isobaric surfaces of $p = 27.5$ and $p = 25.0$ as moving in a corresponding manner, then we see that during this movement the quantities $(E_{27.5}^{25.0})_a$ and $(E_{27.5}^{25.0})_b$, always retain the values 40 200 and 40 100 just calculated for them. Therefore the closed curve, even during its movement, always encloses 100 solenoids. We therefore conclude that the tubular structure in the atmosphere, bounded by vertical walls through the curves 40 200 and 40 100 and by the isobaric surfaces of $p = 27.5$ and $p = 25.0$, encloses exactly 100 unit solenoids whose courses must lie parallel to the curves 40 200 and 40 100. By a series of analogous operations we are led to the conclusion that there are always 100 solenoids between each pair of adjacent curves on the $E_{27.5}^{25.0}$-map (Fig. 12, page 71).

According to Bjerknes' theory these solenoids tend to set up a rotational movement in the atmosphere. The direction of this rotation is expressed by the rule that the air tends to rise where $E_{27.5}^{25.0}$ is large, and to sink where $E_{22.7}^{25.0}$ is small. Thus the movement resulting from the solenoid system of the chart of $E_{27.5}^{25.0}$, page 71, Fig. 12, is an ascending one in the vicinity of Pierre and Topeka, and a descending one in the outer portions of the region shown on the map.

Returning to the closed curve in the atmosphere indicated at ab in Fig. 12, we know first of all that it embraces 100 solenoids. Therefore from the preceding theorem we know that the increase of circulation along this closed curve is at the rate of 100 $\dfrac{mile^2}{hour^2}$ per hour, and that it is directed upward along the vertical a and downward along the vertical b. If this increase in the circulation be divided by the length of the line ab, which from measurement is seen to amount to 125 miles, then, according to the definition of circulation, we obtain a mean tangential acceleration of 0.8 $\dfrac{mile}{hour^2}$ for the air-particles composing the curve. In other words, if we assume that the air was originally at rest, and if we leave out of consideration the influences of friction and the earth's rotation, then this solenoid-system would have produced a mean velocity of 0.8 $\frac{\text{mile}}{\text{hour}}$ along the curve by the end of the first hour. At the end of the second hour the mean velocity would be 1.6 $\frac{\text{mile}}{\text{hour}}$; at the end of the third hour 2.4 $\frac{\text{mile}}{\text{hour}}$ and so on. By carrying out a number of such numerical calculations on the $E_{p_0}^p$-map one will soon become so familiar with the dynamic significance of its lines that a glance at the chart will suffice to recognize and read the accelerations indicated by it.

The $E_{27.5}^{25.0}$-map may be constructed directly from the values telegraphed to the central office; but after the complete results of the kite-observations at the different stations have been received by mail these maps may be constructed for much thinner strata in the atmosphere. Then, for instance, the layer of air between the isobaric surfaces for $p = 27.5$ and $p = 25.0$ can be subdivided into five strata whose dynamic condition can be presented by charts for $E_{27.5}^{27.0}$, $E_{27.0}^{26.5}$, $E_{26.5}^{26.0}$, $E_{26.0}^{25.5}$, $E_{25.5}^{25.0}$. On page 71 only three of these latter maps have been drawn, viz., for $E_{27.5}^{27.0}$ as Fig. 13; $E_{26.5}^{26.0}$ as Fig. 14; and $E_{25.5}^{25.0}$ as Fig. 15. From Fig. 13 it is seen that in the layer of air between the isobaric surfaces of 27.5 and 27.0, the maximum ascensive tendency is southeast of Topeka. The $E_{26.5}^{26.0}$-map, Fig. 14, shows that in the layer between the 26.5 and the 26.0 isobaric surfaces the air has its maximum ascensive tendency just over Topeka. The $E_{25.5}^{25.0}$-map of Fig. 15 shows the maximum ascensive tendency to be above Pierre and Dodge City. If we neglect this shift of the center of ascension toward the northwest then we find that the solenoids as drawn for the thinner strata have nearly the same characters as those drawn for the larger interval of the $E_{27.5}^{25.0}$-map. It suffices, therefore, to construct $E_{p_0}^p$-maps for thicker strata or greater intervals by aid of the telegraphic reports and afterward for smaller intervals by means of the more complete reports by mail. In this way very brief condensed telegraphic reports may be made to do good service.

The general expression for the number of solenoids within a closed curve consisting of two verticals a and b, and two curves lying in the isobaric surfaces $p = p_0$ and $p = p_1$, is

$$A = (E_{p_0}^{p_1})_a - (E_{p_0}^{p_1})_b \qquad (28)$$

This may be deduced in exactly the same way as the special formula equation (27). It follows from equation (28) that each of the tubular-shaped figures in the atmosphere bounded by the isobaric surfaces $p = p_1$ and $p = p_0$, and the vertical walls, passing through the curves drawn on such a map, contains a number of solenoids equal to the number obtained by subtracting the numbers belonging to those latter curves. Hence it follows that in the maps forming Figs. 13, 14 and 15 of page 71 designated as $E_{27.5}^{27.0}$, $E_{26.5}^{26.0}$, $E_{25.5}^{25.0}$, there are always 50 solenoids between each pair of adjacent curved lines.

Significance of $\Pi_{V_0}^{V}$-maps. — The dynamic significance of the charts of $\Pi_{V_0}^{V}$ results from a second principle enunciated by Bjerknes.* If the velocity of the air be indicated by u, and the density of the air by ρ, then $\bar{u} = \rho u$ expresses the amount of the so-called specific quantity of motion of the air. The tangential component u_t of this quantity, when integrated along a closed curve, we call the "moment-circulation" of that curve. By moment-solenoid we designate the tubular figure formed in the atmosphere by surfaces of equal density (isodense surfaces) and by level surfaces of gravity, when these surfaces are constructed for each unit difference of density and of gravity potential, respectively. If we further assume that the barometric formula for static conditions also holds true under dynamic conditions, then Bjerknes' second theorem may be stated somewhat as follows:

The increase, during a unit of time, of the moment-circulation of a closed curve consisting of particles of air in the atmosphere, is equal to the number of moment-solenoids enclosed by that curve. We designate the moment-circulation of the closed curve by $\bar{C}$ and the number of enclosed moment-solenoids by $\bar{A}$; then this theorem is expressed by the formula

$$\frac{d\bar{C}}{dt} = \bar{A} = \int \rho \, dV \qquad (29)$$

where integration is to be carried out along the whole closed curve. Now the numbers and positions of the moment-solenoids are readily determined from the $\Pi_{V_0}^{V}$-maps. To demonstrate this let the closed atmospheric curve be composed of two lines ab and ab, lying in the level-surfaces of gravity $V = V_0$ and $V = V_1$ respectively, and of the two verticals aa and bb, joining the ends of these two lines. Then $\rho \, dV$, and $\int \rho \, dV$, each equal zero along the lines ab in the level surfaces and it only becomes necessary to carry out the integration along the verticals aa and bb.

Thus we obtain

$$\frac{d\bar{C}}{dt} = \left(\int_{V_0}^{V_1} \rho \, dV \right)_a - \left(\int_{V_0}^{V_1} \rho \, dV \right)_b. \qquad (30)$$

By the aid of the static barometric formula

$$dp = -\rho \, dV$$

the above integrals may be transformed. The integration of this formula along the vertical aa gives

$$\left(\int_{V_0}^{V_1} \rho \, dV \right)_a = (p_{V_0} - p_{V_1})_a$$

* V. Bjerknes — "On the formation of circulatory movements and vortices in frictionless fluids."—*Christiania, Videnskabsselskabets Skriften*, 1898, No. 5.

but

$$(p_{V_0} - p_{V_1})_a = (\Pi_{V_0}^{V_1})_a$$

hence

$$\left(\int_{V_0}^{V_1} \rho \, d V \right)_a = (\Pi_{V_0}^{V_1})_a.$$

In an analogous way,

$$\left(\int_{V_0}^{V_1} \rho \, d V \right)_b = (\Pi_{V_0}^{V_1})_b.$$

Then by substitutions in (30) we have

$$\bar{A} = (\Pi_{V_0}^{V_1})_a - (\Pi_{V_0}^{V_1})_b. \tag{31}$$

The number of moment-solenoids, $\bar{A}$, has therefore the same dimensions as a pressure, $i.\ e.$,

$$\frac{\text{mass}}{\text{length, time}^2}$$

or its equivalent,

$$\text{density} \times \frac{\text{length}^2}{\text{time}^2}.$$

If the specific gravity, ρ_0, of water at its maximum density be selected as our unit of density, then

a pressure of 1 inch of the mercurial column $= 16.945 \times \rho_0 \times \dfrac{\text{mile}^2}{\text{hour}^2}.$

If we choose the density of air, ρ_1, at 32° F. and 1 atmosphere of pressure as the unit of density, then

a pressure of 1 inch of the mercurial column $= 13\ 105 \times \rho_1 \times \dfrac{\text{mile}^2}{\text{hour}^2}.$

Finally if $\rho_2 = 0.169\ 45\ \rho_0$ be chosen as unit of density, then

a pressure of 1 inch of the mercurial column $= 100 \times \rho_2 \times \dfrac{\text{mile}^2}{\text{hour}^2}.$

Therefore a closed curve composed of two verticals aa and bb, and two lines, ab and ab lying in the level surfaces $V = V_0$ and $V = V_1$ respectively, for which curve we have $\bar{A} = (\Pi_{V_0}^{V_1})_a - (\Pi_{V_0}^{V_1})_b = 1$ inch of the mercurial barometer column, embraces 16.945 moment-solenoids of the $\rho_0 \cdot \dfrac{\text{mile}^2}{\text{hour}^2}$ -system of dimensions; or 13 105 moment-solenoids of the $\rho_1 \cdot \dfrac{\text{mile}^2}{\text{hour}^2}$ -system, or 100 moment-solenoids of the $\rho_2 \cdot \dfrac{\text{mile}^2}{\text{hour}^2}$ -system.

On the $\Pi_{V_0}^{V_1}$-chart, see pages 67, 69, Figs. 4, 5, 9, curves for each 0.01 inch difference of pressure have been drawn. Therefore each of the tubular figures in the atmosphere bounded by vertical walls through any two adjacent curves of these maps, and by the two level surfaces of gravity, $V = V_0$ and $V = V_1$, embraces

$$0.16945 \; \rho_0 \; \frac{\text{mile}^2}{\text{hour}^2} \; \text{moment-solenoids,}$$

or

$$131.05 \; \rho_1 \; \frac{\text{mile}^2}{\text{hour}^2} \; \text{moment-solenoids,}$$

or

$$1 \times \rho_2 \; \frac{\text{mile}^2}{\text{hour}^2} \; \text{moment-solenoids}$$

according to the standard unit that we adopt.

These moment-solenoids tend to direct the specific quantity of motion upward at places where $\Pi_{V_0}^V$ is small, and downward at points where $\Pi_{V_0}^V$ is large. Accordingly, the air lying between the level surfaces $V = 0$ and $V = 40\,000$ (see Fig. 4) will be pushed upward most strongly in the region northeastward from Omaha. The $\Pi_{40\,000}^{60\,000}$-map (see Fig. 9) shows the greatest ascensive tendency to be over Topeka, and the $\Pi_{40\,000}^{80\,000}$-map (see Fig. 5) shows the greatest upward force to be westward from Pierre.

With the aid of these $\Pi_{V_0}^V$-maps such numerical examples showing the specific quantity of motion can be computed, just as corresponding examples for the velocity were computed from the $E_{p_0}^{p_1}$-maps. It seems to me, however, that from a dynamical point of view the specific quantity of motion is a less convenient quantity than the velocity. Probably the $\Pi_{V_0}^V$-charts will only be used in working on certain special problems, such as the comparison of movements in media of such different densities, as the air and the ocean; or when one wishes to calculate the mass of air transported by the winds.

VI.

CONCLUDING REMARKS.

The connection of the charts that we have here drawn for the higher atmospheric strata with the dynamics of the atmosphere must be clear from the preceding pages. It is to be expected that upon such maps we may easily and naturally present our observations and experience as to atmospheric movements and therefore, it would seem to promise good results if the daily weather-predictions could be based upon such maps. At least this latter is practicable in so far as it would require not more than an hour to work up the data necessary for the telegraphic reports from the kite stations. Certainly within one and a half hours after the descent of the last kite these maps could be drawn and finished at the central office of the Weather Bureau.

On the other hand, practical difficulties will certainly be experienced, through occasional inability to make kite ascensions at the proper hour of the day at a sufficient number of stations. But it is to be expected that better results will be attained as the technique of kite-flying develops. In any case it is very desirable that kite-observations be supplemented by observations of another character. Such supplementary observations are indeed supplied by the measurements of cloud heights and cloud velocities. But in order to utilize these we must make use of the Bjerknes theorem of circulation as perfected by taking into account the earth's rotation. I hope soon to return to the consideration of cloud-observations as supplementary to the high-level charts, and also to the consideration of the importance of such charts in weather prediction.

APPENDIX.

FORMULÆ AND TABLES IN THE METRIC SYSTEM (DATED OCTOBER, 1902).[*]

It is easy to convert the formulæ and tables of the preceding memoir from English into metric measures by using the following relations:

$$\text{Velocity, } 1 \frac{\text{mile}}{\text{hour}} = 0.447\ 032 \frac{\text{meter}}{\text{second}}.$$

$$\text{Acceleration of velocity, } 1 \frac{\text{mile}}{\text{hour}^2} = 0.000\ 124\ 175\ 5 \frac{\text{meter}}{\text{second}^2}.$$

$$\text{Circulation, } 1 \frac{\text{mile}^2}{\text{hour}} = 719.415 \frac{\text{meter}^2}{\text{second}}.$$

$$\text{Acceleration of circulation, } 1 \frac{\text{mile}^2}{\text{hour}^2} = 0.199\ 837\ 5 \frac{\text{meter}^2}{\text{second}^2}.$$

The formulæ of this memoir thus become converted respectively into the following, where the units are the meter, the second of mean solar time and the degree centigrade:

(2) $\quad g = g_0(1 - 0.000\ 000\ 314(z - z_0))$.

(3) $\quad g = g_0(1 + 0.000\ 000\ 196(z_0 - z))$.

(4) $\quad V_0 = g_0 z_0(1 + 0.000\ 000\ 098 z_0)$.

(5) $\quad V = V_0 + g_0 z_1(1 - 0.000\ 000\ 157 z_1)$.

(6) $\quad g_0 = 9.80604(1 - 0.00259 \cos 2\lambda)(1 - 0.000\ 000\ 196 z_0)$.

(17)
$$E_{p_0}^{p_1} = \frac{0.760 \times 9.80604 \times 13.59593}{0.001293052 \times 273} \int_{p_1}^{p_0} T \frac{dp}{p}.$$

(18)
$$E_{p_0}^{p_1} = 660.9 \int_{p_1}^{p_0} (t + 273^\circ) d(\log p).$$

(19)
$$\Pi_{V_0}^{V_1} = p_{V_0} \left(1 - 10^{-\frac{1}{660.9} \int_{V_0}^{V_1} \frac{dV}{t + 273}} \right).$$

(20)
$$t_r - t = \frac{0.377 rf(t + 273)}{p - 0.377 rf}.$$

[*] In order to meet any possible question of priority or responsibility it is proper to say that this memoir by J. W. Sandström was received by Professor Cleveland Abbe in April, 1902, with permission to translate and publish : the appendix in metric measures was received in October, 1902. The translation by Dr. Cleveland Abbe, junior, was finished during 1903, and was read by Professor Abbe at the annual meeting of the American Philosophical Society, April, 1905. — C. A.

(21)
$$E_{p_0}^{p_1} = 660.9 \int_{p_1}^{p_0} (t_r + 273) d(\log p).$$

(22)
$$\Pi_{V_0}^{V_1} = p_{V_0}\left(1 - 10^{-\frac{1}{660.9}\int_{V_0}^{V_1}\frac{dV}{t+273}}\right).$$

(23)
$$E_{p_0}^{p_1} = 660.9(t_r + 273)\log\frac{p_0}{p_1}.$$

(24)
$$\Pi_{V_0}^{V_1} = p_{V_0}\left(1 - 10^{-\frac{V_1 - V_0}{660.9(t_r + 273)}}\right).$$

From formula (23) it results that

$E_{800}^{780} = 7.267\,(t_r + 273^\circ)$	$E_{600}^{580} = 9.731\,(t_r + 273^\circ)$
$E_{780}^{760} = 7.456\,(\ \ '' \ \ )$	$E_{580}^{560} = 10.072\,(\ \ '' \ \ )$
$E_{760}^{740} = 7.654\,(\ \ '' \ \ )$	$E_{560}^{540} = 10.438\,(\ \ '' \ \ )$
$E_{740}^{720} = 7.864\,(\ \ '' \ \ )$	$E_{540}^{520} = 10.832\,(\ \ '' \ \ )$
$E_{720}^{700} = 8.086\,(\ \ '' \ \ )$	$E_{520}^{500} = 11.257\,(\ \ '' \ \ )$
$E_{700}^{680} = 8.320\,(\ \ '' \ \ )$	$E_{500}^{480} = 11.717\,(\ \ '' \ \ )$
$E_{680}^{660} = 8.569\,(\ \ '' \ \ )$	$E_{480}^{460} = 12.216\,(\ \ '' \ \ )$
$E_{660}^{640} = 8.832\,(\ \ '' \ \ )$	$E_{460}^{440} = 12.759\,(\ \ '' \ \ )$
$E_{640}^{620} = 9.113\,(\ \ '' \ \ )$	$E_{440}^{420} = 13.352\,(\ \ '' \ \ )$
$E_{620}^{600} = 9.411\,(t_r + 273^\circ)$	$E_{420}^{400} = 14.004\,(t_r + 273^\circ)$

If we put $V_1 = V_0 + 2\,000$ then from equation (24) we get

$$\Pi_{V_0}^{V_0 + 2\,000} = p_{V_0}\left(1 - 10^{-\frac{2\,000}{660.9(t_r + 273)}}\right).$$

The other equations will not be changed by the introduction of the metric system of measures.

In the following pages are given the most important metric tables, numbered the same as the corresponding tables in English measures in the preceding memoir:

TABLE 1 IN THE METRIC SYSTEM.

$$z_1 (1-0.000000157z_1).$$

z_1 = altitude = 1000	2000	3000	4000	5000	6000	7000	8000	9000	10000 meters
$(V-V_0)/g_0 =$ 999.8	1999.4	2998.6	3997.5	4996.1	5994.4	6992.4	7990.0	8987.3	9984.3

TABLE 4 IN THE METRIC SYSTEM.

THE ACCELERATION OF GRAVITY AT SEALEVEL IN METERS PER SECOND.

Geographic Latitude λ.	0°	1°	2°	3°	4°	5°	6°	7°	8°	9°
0°	9.7806	9.7807	9.7807	9.7808	9.7809	9.7810	9.7812	9.7814	9.7816	9.7819
10	9.7822	9.7825	9.7828	9.7832	9.7836	9.7840	9.7845	9.7850	9.7855	9.7860
20	9.7866	9.7872	9.7878	9.7884	9.7891	9.7897	9.7904	9.7911	9.7919	9.7926
30	9.7934	9.7941	9.7949	9.7957	9.7965	9.7974	9.7982	9.7990	9.7999	9.8008
40	9.8016	9.8025	9.8034	9.8043	9.8052	9.8060	9.8069	9.8078	9.8087	9.8096
50	9.8105	9.8113	9.8122	9.8130	9.8139	9.8147	9.8156	9.8164	9.8172	9.8180
60	9.8187	9.8195	9.8202	9.8210	9.8217	9.8224	9.8230	9.8237	9.8243	9.8249
70	9.8255	9.8261	9.8266	9.8271	9.8276	9.8280	9.8285	9.8288	9.8292	9.8296
80	9.8299	9.8302	9.8305	9.8307	9.8309	9.8311	9.8312	9.8313	9.8314	9.8314

TABLE 5 IN THE METRIC SYSTEM.

DECREASE OF GRAVITY WITH ALTITUDE ABOVE SEALEVEL.

Altitude in meters =	1000	2000	3000	4000	5000	6000	7000	8000	9000	10000
Decrease of g =	−0.0019	−0.0038	−0.0058	−0.0077	−0.0096	−0.0115	−0.0135	−0.0154	−0.0173	−0.0192

TABLE 6 IN THE METRIC SYSTEM.

THE ALTITUDES IN METERS ABOVE SEALEVEL OF THE LEVEL SURFACES OF GRAVITY.

V_0	λ = Geographic Latitude.										V_0
	0°	10°	20°	30°	40°	50°	60°	70°	80°	90°	
	Meter.	Meter.	Meter.	Meter.	Meter.	Meter.	Meter.	Meter.	Meter.	Meter.	
0	0	0	0	0	0	0	0	0	0	0	0
1000	102.2	102.2	102.2	102.1	102.0	101.9	101.8	101.8	101.7	101.7	1000
2000	204.5	204.5	204.4	204.2	204.1	203.9	203.7	203.6	203.5	203.4	2000
3000	306.7	306.7	306.6	306.3	306.1	305.8	305.5	305.3	305.2	305.2	3000
4000	409.0	408.9	408.7	408.4	408.1	407.7	407.4	407.1	406.9	406.9	4000
5000	511.2	511.1	510.9	510.6	510.1	509.7	509.2	508.9	508.7	508.6	5000
6000	613.5	613.4	613.1	612.7	612.2	611.6	611.1	610.7	610.4	610.3	6000
7000	715.8	715.7	715.3	714.8	714.2	713.6	712.9	712.5	712.2	712.0	7000
8000	818.0	817.9	817.5	816.9	816.3	815.5	814.8	814.3	813.9	813.8	8000
9000	920.3	920.2	919.7	919.1	918.3	917.4	916.7	916.1	915.6	915.5	9000
10000	1022.5	1022.4	1021.9	1021.2	1020.3	1019.4	1018.6	1017.9	1017.4	1017.3	10000
11000	1124.8	1124.6	1124.1	1123.3	1122.4	1121.3	1120.4	1119.7	1119.2	1119.0	11000
12000	1227.1	1226.9	1226.3	1225.5	1224.4	1223.3	1222.3	1221.5	1220.9	1220.7	12000
13000	1329.3	1329.1	1328.5	1327.6	1326.5	1325.3	1324.2	1323.3	1322.7	1322.5	13000
14000	1431.6	1431.4	1430.7	1429.7	1428.5	1427.2	1426.0	1425.1	1424.4	1424.2	14000
15000	1533.9	1533.6	1532.9	1531.9	1530.6	1529.2	1527.9	1526.9	1526.2	1526.0	15000
16000	1636.1	1635.9	1635.1	1634.0	1632 6	1631.2	1629.8	1628.7	1627.9	1627.7	16000
17000	1738.4	1738.1	1737.4	1736.1	1734.7	1733.1	1731.7	1730.5	1729.7	1729.4	17000
18000	1840.7	1840.4	1839.6	1838.3	1836.8	1835.1	1833.6	1832.3	1831.4	1831.2	18000
19000	1943.0	1942.7	1941.8	1940.4	1938.8	1937.1	1935.5	1934.1	1933.2	1932.9	19000
20000	2045.3	2044.9	2044.0	2042.6	2040.9	2039.0	2037.3	2035.9	2035.0	2034.7	20000
21000	2147.6	2147.2	2146.2	2144.7	2143.0	2141.0	2139.2	2137.7	2136.8	2136.5	21000
22000	2249.9	2249.5	2248.5	2246.9	2245.0	2243.0	2241.1	2239.6	2238.6	2238.2	22000
23000	2352.1	2351.8	2350.7	2349.1	2347.1	2345.0	2343.0	2341.4	2340.3	2340.0	23000
24000	2454.4	2454.0	2452.9	2451.2	2449.2	2446.9	2444.9	2443.2	2442.1	2441.7	24000
25000	2556.7	2556.3	2555.1	2553.4	2551.2	2548.9	2546.8	2545.0	2543.9	2543.5	25000
26000	2659.0	2658.6	2657.4	2655.5	2653.3	2650.9	2648.7	2646.9	2645.7	2645.3	26000
27000	2761.3	2760.9	2759.6	2757.7	2755.4	2752.9	2750.6	2748.7	2747.5	2747.0	27000
28000	2863.6	2863.1	2861.9	2859.9	2857.5	2854.9	2852.5	2850.5	2849.3	2848.8	28000
29000	2965.9	2965.4	2964.1	2962.0	2959.6	2956.9	2954.4	2952.4	2951.0	2950.6	29000
30000	3068.2	3067.6	3066.3	3064.2	3061.6	3058.9	3056.3	3054.2	3052.8	3052.4	30000

TABLE 7 IN THE METRIC SYSTEM.

THE VALUES OF $t_1 - t$ FOR ANY PRESSURE AND TEMPERATURE AND SATURATED AIR.

Barometric Pressure in mm.

Temp. Cent. t	400	420	440	460	480	500	520	540	560	580	600	620	640	660	680	700	720	740	760	780	800	Temp. Cent. t
−30°	0.1	0.1	0.1	0.1	0.1	0.1	0.1	0.1	0.1	0.1	0.1	0.1	0.1	0.1	0.1	0.0	0.0	0.0	0.0	0.0	0.0	−30°
−20	0.2	0.2	0.2	0.2	0.2	0.2	0.2	0.2	0.2	0.2	0.2	0.2	0.2	0.2	0.1	0.1	0.1	0.1	0.1	0.1	0.1	−20
−15	0.4	0.3	0.3	0.3	0.3	0.3	0.3	0.3	0.3	0.3	0.3	0.2	0.2	0.2	0.2	0.2	0.2	0.2	0.2	0.2	0.2	−15
−10	0.5	0.5	0.5	0.5	0.4	0.4	0.4	0.4	0.4	0.4	0.4	0.4	0.3	0.3	0.3	0.3	0.3	0.3	0.2	0.2	0.2	−10
− 8	0.6	0.6	0.6	0.6	0.5	0.5	0.5	0.5	0.5	0.4	0.4	0.4	0.4	0.4	0.4	0.4	0.4	0.3	0.3	0.3	0.3	− 8
− 6	0.7	0.7	0.7	0.7	0.6	0.6	0.6	0.6	0.5	0.5	0.5	0.5	0.5	0.4	0.4	0.4	0.4	0.4	0.4	0.3	0.3	− 6
− 4	0.9	0.8	0.8	0.8	0.7	0.7	0.7	0.6	0.6	0.6	0.6	0.6	0.5	0.5	0.5	0.5	0.5	0.5	0.5	0.5	0.4	− 4
− 2	1.0	1.0	0.9	0.9	0.8	0.8	0.8	0.7	0.7	0.7	0.7	0.6	0.6	0.6	0.6	0.6	0.6	0.6	0.5	0.5	0.5	− 2
0	1.2	1.1	1.1	1.0	1.0	0.9	0.9	0.9	0.8	0.8	0.8	0.8	0.7	0.7	0.7	0.7	0.7	0.6	0.6	0.6	0.6	0
1	1.3	1.2	1.2	1.1	1.1	1.0	1.0	0.9	0.9	0.9	0.9	0.8	0.8	0.8	0.8	0.7	0.7	0.7	0.7	0.6	0.6	1
2	1.4	1.3	1.2	1.2	1.1	1.1	1.1	1.0	1.0	0.9	0.9	0.9	0.9	0.8	0.8	0.8	0.8	0.7	0.7	0.7	0.7	2
3	1.5	1.4	1.3	1.3	1.2	1.2	1.1	1.1	1.1	1.1	1.0	1.0	1.0	0.9	0.9	0.9	0.8	0.8	0.8	0.8	0.7	3
4	1.6	1.5	1.5	1.4	1.3	1.3	1.2	1.2	1.1	1.1	1.1	1.0	1.0	1.0	0.9	0.9	0.9	0.9	0.8	0.8	0.8	4
5	1.7	1.6	1.6	1.5	1.4	1.4	1.3	1.3	1.2	1.2	1.1	1.1	1.1	1.0	1.0	1.0	1.0	0.9	0.9	0.9	0.8	5
6	1.8	1.7	1.7	1.6	1.5	1.5	1.4	1.4	1.3	1.3	1.2	1.2	1.2	1.1	1.1	1.1	1.0	1.0	1.0	0.9	0.9	6
7	2.0	1.9	1.8	1.7	1.7	1.6	1.5	1.5	1.4	1.4	1.3	1.3	1.2	1.2	1.2	1.1	1.1	1.1	1.0	1.0	1.0	7
8	2.1	2.0	1.9	1.9	1.8	1.7	1.6	1.6	1.5	1.5	1.4	1.4	1.3	1.3	1.3	1.2	1.2	1.1	1.1	1.1	1.0	8
9	2.3	2.2	2.1	2.0	1.9	1.8	1.8	1.7	1.6	1.6	1.5	1.5	1.4	1.4	1.3	1.3	1.3	1.2	1.2	1.2	1.1	9
10	2.5	2.3	2.2	2.1	2.0	2.0	1.9	1.8	1.7	1.7	1.6	1.6	1.5	1.5	1.4	1.4	1.4	1.3	1.3	1.3	1.2	10
11	2.6	2.5	2.4	2.3	2.2	2.1	2.0	1.9	1.9	1.8	1.8	1.7	1.6	1.6	1.5	1.5	1.5	1.4	1.4	1.3	1.3	11
12	2.8	2.7	2.6	2.5	2.4	2.3	2.2	2.1	2.0	1.9	1.9	1.8	1.8	1.7	1.7	1.6	1.6	1.5	1.5	1.4	1.4	12
13	3.0	2.9	2.8	2.6	2.5	2.4	2.3	2.2	2.2	2.1	2.0	1.9	1.9	1.8	1.8	1.7	1.7	1.6	1.6	1.5	1.5	13
14	3.3	3.1	3.0	2.8	2.7	2.6	2.5	2.4	2.3	2.2	2.2	2.1	2.0	2.0	1.9	1.8	1.8	1.7	1.7	1.7	1.6	14
15	3.5	3.3	3.2	3.0	2.9	2.8	2.7	2.6	2.5	2.4	2.3	2.2	2.1	2.0	2.0	2.0	1.9	1.9	1.8	1.8	1.7	15
16	3.7	3.5	3.4	3.2	3.1	3.0	2.9	2.8	2.7	2.6	2.5	2.4	2.3	2.2	2.2	2.1	2.1	2.0	2.0	1.9	1.8	16
17	4.0	3.8	3.6	3.5	3.3	3.2	3.1	3.0	2.8	2.7	2.6	2.6	2.5	2.4	2.3	2.3	2.2	2.1	2.1	2.0	2.0	17
18	4.3	4.1	3.9	3.7	3.5	3.4	3.3	3.2	3.0	2.9	2.8	2.7	2.7	2.6	2.6	2.5	2.4	2.4	2.3	2.2	2.1	18
19	4.6	4.3	4.1	4.0	3.8	3.6	3.5	3.4	3.2	3.1	3.0	2.9	2.8	2.8	2.7	2.6	2.5	2.4	2.4	2.3	2.2	19
20	4.9	4.6	4.4	4.2	4.1	3.9	3.7	3.6	3.5	3.3	3.2	3.1	3.0	2.9	2.9	2.8	2.7	2.6	2.5	2.5	2.4	20
21	5.2	5.0	4.7	4.5	4.3	4.2	4.0	3.8	3.7	3.6	3.5	3.3	3.2	3.1	3.0	2.9	2.8	2.7	2.6	2.6	2.5	21
22	5.6	5.3	5.0	4.8	4.6	4.4	4.3	4.1	3.9	3.8	3.7	3.6	3.4	3.3	3.2	3.2	3.1	3.0	2.9	2.8	2.7	22
23	5.9	5.7	5.4	5.1	4.9	4.7	4.5	4.4	4.2	4.1	3.9	3.8	3.7	3.6	3.5	3.4	3.3	3.2	3.1	3.0	2.9	23
24			5.7	5.5	5.2	5.0	4.9	4.7	4.5	4.3	4.2	4.0	3.9	3.8	3.8	3.7	3.6	3.5	3.4	3.3	3.1	24
25				5.9	5.6	5.4	5.2	5.0	4.8	4.6	4.5	4.3	4.2	4.1	3.9	3.8	3.7	3.6	3.5	3.4	3.3	25
26						5.7	5.5	5.3	5.1	4.9	4.8	4.6	4.5	4.3	4.2	4.1	4.0	3.8	3.7	3.6	3.5	26
27							5.9	5.6	5.4	5.3	5.1	4.9	4.7	4.6	4.5	4.3	4.2	4.1	4.0	3.9	3.8	27
28									5.8	5.6	5.4	5.2	5.1	4.9	4.8	4.6	4.5	4.4	4.3	4.1	4.0	28
29										5.9	5.7	5.6	5.4	5.2	5.1	4.9	4.8	4.6	4.5	4.4	4.3	29
30												5.9	5.7	5.5	5.4	5.2	5.1	4.9	4.8	4.7	4.5	30
31														5.9	5.7	5.6	5.4	5.2	5.1	5.0	4.8	31
32																5.9	5.7	5.6	5.4	5.3	5.2	32
33																		5.9	5.8	5.6	5.5	33
34																				6.0	5.8	34
	400	420	440	460	480	500	520	540	560	580	600	620	640	660	680	700	720	740	760	780	800	

TABLE 8 IN THE METRIC SYSTEM.

THE VALUES OF $t_r - t$ FOR ANY VALUE OF $t_1 - t$ AND RELATIVE HUMIDITY.

Relative Humidity.	$t_1-t.$															Relative Humidity.
	0.2	0.4	0.6	0.8	1.0	1.2	1.4	1.6	1.8	2.0	2.2	2.4	2.6	2.8	3.0	
10%	0.0	0.0	0.1	0.1	0.1	0.1	0.1	0.2	0.2	0.2	0.2	0.2	0.3	0.3	0.3	10%
20	0.0	0.1	0.1	0.2	0.2	0.2	0.3	0.3	0.4	0.4	0.4	0.5	0.5	0.6	0.6	20
30	0.1	0.1	0.2	0.2	0.3	0.4	0.4	0.5	0.5	0.6	0.7	0.7	0.8	0.8	0.9	30
40	0.1	0.2	0.2	0.3	0.4	0.5	0.6	0.6	0.7	0.8	0.9	1.0	1.0	1.1	1.2	40
50	0.1	0.2	0.3	0.4	0.5	0.6	0.7	0.8	0.9	1.0	1.1	1.2	1.3	1.4	1.5	50
60	0.1	0.2	0.4	0.5	0.6	0.7	0.8	1.0	1.1	1.2	1.3	1.4	1.6	1.7	1.8	60
70	0.1	0.3	0.4	0.6	0.7	0.8	1.0	1.1	1.3	1.4	1.5	1.7	1.8	2.0	2.1	70
80	0.2	0.3	0.5	0.6	0.8	1.0	1.1	1.3	1.4	1.6	1.8	1.9	2.1	2.2	2.4	80
90	0.2	0.4	0.5	0.7	0.9	1.1	1.3	1.4	1.6	1.8	2.0	2.2	2.3	2.5	2.7	90
100	0.2	0.4	0.6	0.8	1.0	1.2	1.4	1.6	1.8	2.0	2.2	2.4	2.6	2.8	3.0	100

Relative Humidity.	$t_1-t.$															Relative Humidity.
	3.2	3.4	3.6	3.8	4.0	4.2	4.4	4.6	4.8	5.0	5.2	5.4	5.6	5.8	6.0	
10%	0.3	0.3	0.4	0.4	0.4	0.4	0.4	0.5	0.5	0.5	0.5	0.5	0.6	0.6	0.6	10%
20	0.6	0.7	0.7	0.8	0.8	0.8	0.9	0.9	1.0	1.0	1.0	1.1	1.1	1.2	1.2	20
30	1.0	1.0	1.1	1.1	1.2	1.3	1.3	1.4	1.4	1.5	1.6	1.6	1.7	1.7	1.8	30
40	1.3	1.4	1.4	1.5	1.6	1.7	1.8	1.8	1.9	2.0	2.1	2.2	2.2	2.3	2.4	40
50	1.6	1.7	1.8	1.9	2.0	.2.1	2.2	2.3	2.4	2.5	2.6	2.7	2.8	2.9	3.0	50
60	1.9	2.0	2.2	2.3	2.4	2.5	2.6	2.8	2.9	3.0	3.1	3.2	3.4	3.5	3.6	60
70	2.2	2.4	2.5	2.7	2.8	2.9	3.1	3.2	3.4	3.5	3.6	3.8	3.9	4.1	4.2	70
80	2.6	2.7	2.9	3.0	3.2	3.4	3.5	3.7	3.8	4.0	4.2	4.3	4.5	4.6	4.8	80
90	2.9	3.1	3.2	3.4	3.6	3.8	4.0	4.1	4.3	4.5	4.7	4.9	5.0	5.2	5.4	90
100	3.2	3.4	3.6	3.8	4.0	4.2	4.4	4.6	4.8	5.0	5.2	5.4	5.6	5.8	6.0	100

CONSTRUCTION OF ISOBARIC CHARTS

TABLE 9 IN THE METRIC SYSTEM.

THE VALUES OF $E^{p_1}_{p_0}$.

t_r	E^{780}_{800}	E^{760}_{780}	E^{740}_{760}	E^{720}_{740}	E^{700}_{720}	E^{680}_{700}	E^{660}_{680}	E^{640}_{660}	E^{620}_{640}	E^{600}_{620}	E^{580}_{600}	E^{560}_{580}	E^{540}_{560}	E^{520}_{540}	E^{500}_{520}	E^{480}_{500}	E^{460}_{480}	E^{440}_{460}	E^{420}_{440}	E^{400}_{420}	t_r
°C.																					°C.
—20	1839	1886	1936	1990	2046	2105	2168	2234	2306	2381	2462	2548	2641	2740	2848	2964	3091	3228	3378	3543	—20
—19	1846	1894	1944	1997	2054	2113	2177	2243	2315	2390	2472	2558	2651	2751	2859	2976	3103	3241	3391	3557	—19
—18	1853	1901	1952	2005	2062	2122	2185	2252	2324	2400	2481	2568	2662	2762	2871	2988	3115	3254	3405	3571	—18
—17	1860	1909	1959	2013	2070	2130	2194	2261	2333	2409	2491	2578	2672	2773	2882	3000	3127	3266	3418	3585	—17
—16	1868	1916	1967	2021	2078	2138	2202	2270	2342	2419	2501	2589	2683	2784	2893	3011	3140	3279	3431	3599	—16
—15	1875	1924	1975	2029	2086	2147	2211	2279	2351	2428	2511	2599	2693	2795	2904	3023	3152	3292	3445	3613	—15
—14	1882	1931	1982	2037	2094	2155	2219	2287	2360	2437	2520	2609	2703	2805	2916	3035	3164	3305	3458	3627	—14
—13	1889	1939	1990	2045	2102	2163	2228	2296	2369	2447	2530	2619	2714	2816	2927	3046	3176	3317	3472	3641	—13
—12	1897	1946	1998	2053	2110	2172	2237	2305	2378	2456	2540	2629	2724	2827	2938	3058	3188	3330	3485	3655	—12
—11	1904	1953	2005	2060	2119	2180	2245	2314	2388	2466	2550	2639	2735	2838	2949	3070	3201	3343	3498	3669	—11
—10	1911	1961	2013	2068	2127	2188	2254	2323	2397	2475	2559	2649	2745	2849	2961	3082	3213	3356	3512	3683	—10
— 9	1918	1968	2021	2076	2135	2196	2262	2332	2406	2485	2569	2659	2756	2860	2972	3093	3225	3368	3525	3697	— 9
— 8	1926	1976	2028	2084	2143	2205	2271	2340	2415	2494	2579	2669	2766	2870	2983	3105	3237	3381	3538	3711	— 8
— 7	1933	1983	2036	2092	2151	2213	2279	2349	2424	2503	2588	2679	2777	2881	2994	3117	3249	3394	3552	3725	— 7
— 6	1940	1991	2044	2100	2159	2221	2288	2358	2433	2513	2598	2689	2787	2892	3006	3128	3262	3407	3565	3739	— 6
— 5	1948	1998	2051	2108	2167	2230	2296	2367	2442	2522	2608	2699	2797	2903	3017	3140	3274	3419	3578	3753	— 5
— 4	1955	2006	2059	2115	2175	2238	2305	2376	2451	2532	2618	2709	2808	2914	3028	3152	3286	3432	3592	3767	— 4
— 3	1962	2013	2067	2123	2183	2246	2314	2385	2461	2541	2627	2719	2818	2925	3039	3164	3298	3445	3605	3781	— 3
— 2	1969	2021	2074	2131	2191	2255	2322	2393	2470	2550	2637	2730	2829	2935	3051	3175	3311	3458	3618	3795	— 2
— 1	1977	2028	2082	2139	2199	2263	2331	2402	2479	2560	2647	2740	2839	2946	3062	3187	3323	3470	3632	3809	— 1
0	1984	2035	2090	2147	2207	2271	2339	2411	2488	2569	2657	2750	2850	2957	3073	3199	3335	3483	3645	3823	0
1	1991	2043	2097	2155	2216	2280	2348	2420	2497	2579	2666	2760	2860	2968	3084	3210	3347	3496	3658	3837	1
2	1998	2050	2105	2163	2224	2288	2356	2429	2506	2588	2676	2770	2870	2979	3096	3222	3359	3509	3672	3851	2
3	2006	2058	2113	2170	2232	2296	2365	2438	2515	2597	2686	2780	2881	2990	3107	3234	3372	3521	3685	3865	3
4	2013	2065	2120	2178	2240	2305	2374	2446	2524	2607	2696	2790	2891	3000	3118	3246	3384	3534	3699	3879	4
5	2020	2073	2128	2186	2248	2313	2382	2455	2533	2616	2705	2800	2902	3011	3129	3257	3396	3547	3712	3893	5
6	2027	2080	2135	2194	2256	2321	2391	2464	2543	2626	2715	2810	2912	3022	3141	3269	3408	3560	3725	3907	6
7	2035	2088	2143	2202	2264	2330	2399	2473	2552	2635	2725	2820	2923	3033	3152	3281	3420	3573	3739	3921	7
8	2042	2095	2151	2210	2272	2338	2408	2482	2561	2644	2734	2830	2933	3044	3163	3292	3433	3585	3752	3935	8
9	2049	2103	2158	2218	2280	2346	2416	2491	2570	2654	2744	2840	2944	3055	3174	3304	3445	3598	3765	3949	9
10	2057	2110	2166	2226	2288	2355	2425	2499	2579	2663	2754	2850	2954	3065	3186	3316	3457	3611	3779	3963	10
11	2064	2118	2174	2233	2296	2363	2434	2508	2588	2673	2764	2860	2964	3076	3197	3328	3469	3624	3792	3977	11
12	2071	2125	2181	2241	2305	2371	2442	2517	2597	2682	2773	2871	2975	3087	3208	3339	3482	3636	3805	3991	12
13	2078	2132	2189	2249	2313	2380	2451	2526	2606	2692	2783	2881	2985	3098	3220	3351	3494	3649	3819	4005	13
14	2086	2140	2197	2257	2321	2388	2459	2535	2615	2701	2793	2891	2996	3109	3231	3363	3506	3662	3832	4019	14
15	2093	2147	2204	2265	2329	2396	2468	2544	2625	2710	2803	2901	3006	3120	3242	3374	3518	3675	3845	4033	15
16	2100	2155	2212	2273	2337	2404	2476	2552	2634	2720	2812	2911	3017	3130	3253	3386	3530	3687	3859	4047	16
17	2107	2162	2220	2281	2345	2413	2485	2561	2643	2729	2822	2921	3027	3141	3265	3398	3543	3700	3872	4061	17
18	2115	2170	2227	2288	2353	2421	2494	2570	2652	2739	2832	2931	3037	3152	3276	3410	3555	3713	3885	4075	18
19	2122	2177	2235	2296	2361	2429	2502	2579	2661	2748	2841	2941	3048	3163	3287	3421	3567	3726	3899	4089	19
20	2129	2185	2243	2304	2369	2438	2511	2588	2670	2757	2851	2951	3058	3174	3298	3433	3578	3738	3912	4103	20
21	2136	2192	2250	2312	2377	2446	2519	2597	2679	2767	2861	2961	3069	3185	3310	3445	3592	3751	3925	4117	21
22	2144	2200	2258	2320	2385	2454	2528	2605	2688	2776	2871	2971	3079	3195	3321	3457	3604	3764	3939	4131	22
23	2151	2207	2266	2328	2393	2463	2536	2614	2697	2786	2880	2981	3090	3206	3332	3468	3616	3777	3952	4145	23
24	2158	2214	2273	2336	2402	2471	2545	2623	2707	2795	2890	2991	3100	3217	3343	3480	3628	3789	3966	4159	24
25	2166	2222	2281	2343	2410	2479	2554	2632	2716	2804	2900	3001	3111	3228	3355						25
26	2173	2229	2289	2351	2418	2488	2562	2641	2725	2814	2910	3012	3121	3239	3366						26
27	2180	2237	2296	2359	2426	2496	2571	2650	2734	2823	2919	3022	3131	3250	3377						27
28	2187	2244	2304	2367	2434	2504	2579	2658	2743	2833	2929	3032	3142	3260	3388						28
29	2195	2252	2312	2375	2442	2513	2588	2667	2752	2842	2939	3042	3152	3271	3400						29
30	2202	2259	2319	2383	2450	2521	2596	2676	2761	2852											30
31	2209	2267	2327	2391	2458	2529	2605	2685	2770	2861											31
32	2216	2274	2334	2399	2466	2538	2614	2694	2779	2870											32
33	2224	2282	2342	2406	2474	2546	2622	2703	2789	2880											33
34	2231	2289	2350	2414	2482	2554	2631	2711	2798	2889											34
35	2238	2296	2357	2422	2490																35
36	2246	2304	2365	2430	2499																36
37	2253	2311	2373	2438	2507																37
38	2260	2319	2380	2446	2515																38
39	2762	2326	2388	2454	2523																39
	E^{780}_{800}	E^{760}_{780}	E^{740}_{760}	E^{720}_{740}	E^{700}_{720}	E^{680}_{700}	E^{660}_{680}	E^{640}_{660}	E^{620}_{640}	E^{600}_{620}	E^{580}_{600}	E^{560}_{580}	E^{540}_{560}	E^{520}_{540}	E^{500}_{520}	E^{480}_{500}	E^{460}_{480}	E^{440}_{460}	E^{420}_{440}	E^{400}_{420}	

TABLE 10 IN THE METRIC SYSTEM.

THE VALUES OF $\Pi_{V_0}^{V_0+2000}$.

p_γ	Virtual Temperature t_γ Centigrade.													p_γ
	−20°	−15°	−10°	−5°	0°	5°	10°	15°	20°	25°	30°	35°	40°	
mm.														mm.
400	10.87	10.66	10.46	10.27	10.08	9.90	9.73	9.56						400
410	11.14	10.93	10.72	10.52	10.33	10.15	9.97	9.80						410
420	11.41	11.19	10.98	10.78	10.58	10.40	10.21	10.04						420
430	11.68	11.46	11.24	11.04	10.84	10.64	10.46	10.28						430
440	11.95	11.72	11.50	11.29	11.09	10.89	10.70	10.52						440
450	12.22	11.99	11.77	11.55	11.34	11.14	10.94	10.76	10.58					450
460	12.50	12.26	12.03	11.81	11.59	11.39	11.19	11.00	10.81					460
470	12.77	12.52	12.29	12.06	11.84	11.63	11.43	11.23	11.05					470
480	13.04	12.79	12.55	12.32	12.10	11.88	11.67	11.47	11.28					480
490	13.31	13.06	12.81	12.58	12.35	12.13	11.92	11.71	11.52					490
500	13.58	13.32	13.07	12.83	12.60	12.38	12.16	11.95	11.75	11.56				500
510	13.85	13.59	13.33	13.09	12.85	12.62	12.40	12.19	11.99	11.79				510
520	14.13	13.86	13.60	13.35	13.10	12.87	12.65	12.43	12.22	12.02				520
530	14.40	14.12	13.86	13.60	13.36	13.12	12.89	12.67	12.46	12.25				530
540	14.67	14.39	14.12	13.86	13.61	13.37	13.13	12.91	12.69	12.48				540
550	14.94	14.66	14.38	14.12	13.86	13.61	13.38	13.15	12.93	12.71	12.50			550
560	15.21	14.92	14.64	14.37	14.11	13.86	13.62	13.39	13.16	12.94	12.73			560
570	15.48	15.19	14.90	14.63	14.36	14.11	13.86	13.63	13.40	13.17	12.96			570
580	15.76	15.46	15.17	14.89	14.62	14.36	14.11	13.86	13.63	13.40	13.18			580
590	16.03	15.72	15.43	15.14	14.87	14.60	14.35	14.10	13.87	13.64	13.41			590
600	16.30	15.99	15.69	15.40	15.12	14.85	14.59	14.34	14.10	13.87	13.64	13.42		600
610	16.57	16.25	15.95	15.66	15.37	15.10	14.84	14.58	14.34	14.10	13.87	13.65		610
620	16.84	16.52	16.21	15.91	15.62	15.35	15.08	14.82	14.57	14.33	14.10	13.87		620
630	17.11	16.79	16.47	16.17	15.88	15.59	15.32	15.06	14.81	14.56	14.32	14.09		630
640	17.39	17.05	16.73	16.43	16.13	15.84	15.57	15.30	15.04	14.79	14.55	14.32		640
650	17.66	17.32	17.00	16.68	16.38	16.09	15.81	15.54	15.28	15.02	14.78	14.54	14.31	650
660	17.93	17.59	17.26	16.94	16.63	16.34	16.05	15.78	15.51	15.25	15.00	14.76	14.53	660
670	18.20	17.85	17.52	17.20	16.88	16.58	16.30	16.02	15.75	15.48	15.23	14.99	14.75	670
680	18.47	18.12	17.78	17.45	17.14	16.83	16.54	16.25	15.98	15.72	15.46	15.21	14.97	680
690	18.74	18.39	18.04	17.71	17.39	17.08	16.78	16.49	16.22	15.95	15.69	15.43	15.19	690
700	19.02	18.65	18.30	17.97	17.64	17.33	17.02	16.73	16.45	16.18	15.91	15.66	15.41	700
710	19.27	18.92	18.56	18.22	17.89	17.57	17.27	16.97	16.69	16.41	16.14	15.88	15.63	710
720	19.56	19.19	18.83	18.48	18.14	17.82	17.51	17.21	16.92	16.64	16.37	16.11	15.85	720
730	19.83	19.45	19.09	18.74	18.40	18.07	17.75	17.45	17.16	16.87	16.60	16.33	16.07	730
740	20.10	19.72	19.35	18.99	18.65	18.32	18.00	17.69	17.39	17.10	16.82	16.55	16.29	740
750	20.37	19.99	19.61	19.25	18.90	18.56	18.24	17.93	17.63	17.33	17.05	16.78	16.51	750
760	20.65	20.25	19.87	19.51	19.15	18.81	18.48	18.17	17.86	17.56	17.28	17.00	16.73	760
770	20.92	20.52	20.13	19.76	19.40	19.06	18.73	18.41	18.10	17.80	17.51	17.22	16.95	770
780	21.19	20.78	20.39	20.02	19.66	19.31	18.97	18.65	18.33	18.03	17.73	17.45	17.17	780
790	21.46	21.05	20.66	20.28	19.91	19.55	19.21	18.88	18.57	18.26	17.96	17.67	17.39	790
	−20°	−15°	−10°	−5°	0°	5°	10°	15°	20°	25°	30°	35°	40°	

TABLE 11 IN THE METRIC SYSTEM.

VALUES OF Π_{3707}^{4000} FOR OMAHA; WHERE $\lambda = 41°\ 16'$ N.; $z_0 = 378.2$ METERS, $g_0 = 9.8020$ MET.²/SEC.² WHENCE $V_0 = 3\ 707$ MET.²/SEC.².

p_{3707}	t_γ = Virtual Temperature.							p_{3707}
	−20° C.	−10° C.	0° C.	10° C.	20° C.	30° C.	40° C.	
680	2.74	2.63	2.54	2.45	2.37	2.28	2.21	680
690	2.78	2.67	2.57	2.48	2.40	2.32	2.24	690
700	2.82	2.71	2.61	2.52	2.44	2.35	2.28	700
710	2.86	2.75	2.65	2.56	2.47	2.39	2.31	710
720	2.90	2.79	2.69	2.59	2.51	2.42	2.34	720
730	2.94	2.83	2.72	2.63	2.54	2.45	2.37	730
740	2.98	2.86	2.76	2.66	2.58	2.49	2.41	740
750	3.02	2.90	2.80	2.70	2.61	2.52	2.44	750
760	3.06	2.94	2.83	2.74	2.64	2.55	2.47	760
770	3.10	2.98	2.87	2.77	2.68	2.59	2.50	770

TABLE 12 IN THE METRIC SYSTEM.

THE VALUES OF $\left(E_{p_0}^{p_1}\right)_{(t_r\,=\,0°\,C.)}$, OR THE NUMBER OF LEVEL SURFACES BETWEEN p_0 AND p_1 WHEN $t_r = 0°$ C.

p_1	p_0	0	1	2	3	4	5	6	7	8	9
mm. 400	mm. 400	0	196	391	585	780	973	1167	1359	1552	1744
	410	1935	2126	2316	2506	2696	2885	3073	3261	3449	3636
420	420	0	186	372	558	743	927	1111	1295	1478	1661
	430	1843	2025	2207	2389	2570	2750	2930	3109	3288	3467
440	440	0	178	355	532	709	885	1061	1237	1412	1587
	450	1761	1935	2108	2281	2454	2627	2799	2971	3142	3313
460	460	0	170	340	509	678	847	1015	1183	1351	1518
	470	1685	1852	2018	2184	2349	2514	2679	2844	3008	3172
480	480	0	163	326	488	650	812	973	1134	1295	1456
	490	1616	1775	1935	2094	2253	2411	2569	2727	2885	3042
500	500	0	157	313	469	624	779	934	1089	1244	1398
	510	1552	1705	1858	2011	2164	2316	2468	2620	2771	2922
520	520	0	150	301	451	600	750	899	1048	1196	1345
	530	1493	1640	1788	1935	2082	2228	2375	2521	2667	2812
540	540	0	145	290	434	578	722	866	1009	1152	1295
	550	1438	1580	1722	1864	2006	2147	2288	2429	2569	2710
560	560	0	140	279	419	558	696	835	973	1111	1249
	570	1387	1524	1661	1798	1935	2071	2207	2343	2479	2614
580	580	0	135	270	404	539	673	807	940	1073	1206
	590	1339	1472	1605	1737	1869	2001	2132	2264	2395	2526
600	600	0	130	261	391	521	650	780	909	1038	1167
	610	1295	1424	1552	1680	1807	1935	2062	2189	2316	2443
620	620	0	126	252	378	504	629	755	880	1005	1129
	630	1254	1378	1502	1626	1750	1873	1996	2119	2242	2365
640	640	0	122	244	366	488	610	731	852	973	1094
	650	1215	1335	1456	1576	1696	1815	1935	2054	2173	2292
660	660	0	119	237	355	474	592	709	827	944	1061
	670	1178	1295	1412	1528	1645	1761	1877	1992	2108	2224
680	680	0	115	230	345	460	574	688	802	916	1030
	690	1144	1257	1371	1484	1597	1710	1822	1935	2047	2159
700	700	0	112	224	335	446	558	669	780	890	1001
	710	1111	1222	1332	1442	1552	1661	1771	1880	1989	2099
720	720	0	109	217	326	434	542	650	758	866	973
	730	1081	1188	1295	1402	1509	1616	1722	1829	1935	2041
740	740	0	106	211	317	422	528	633	738	843	947
	750	1052	1156	1260	1365	1469	1572	1676	1780	1883	1987
760	760	0	103	206	309	411	514	616	718	821	923
	770	1024	1126	1228	1329	1430	1531	1632	1733	1834	1935
780	780	0	100	201	301	401	501	601	700	800	899
	790	998	1097	1196	1295	1394	1493	1591	1689	1788	1886

TABLE 13 IN THE METRIC SYSTEM.

THE VALUES OF $\frac{t_r}{273} \cdot \left(E_{p_0}^{p_1}\right)_{(t_r=0° C.)}$

$E_{p_0(t_r=0° C.)}^{p_1}$	t_r													$E_{p_0(t_r=0° C.)}^{p_1}$
	-20	-15	-10	-5	0	5	10	15	20	25	30	35	40	
0	0	0	0	0	0	0	0	0	0	0	0	0	0	0
100	-7	-5	-4	-2	0	2	4	5	7	9	11	13	15	100
200	15	11	7	4	0	4	7	11	15	18	22	26	29	200
300	22	16	11	5	0	5	11	16	22	27	33	38	44	300
400	29	22	15	7	0	7	15	22	29	37	44	51	59	400
500	-37	-27	-18	-9	0	9	18	27	37	46	55	64	73	500
600	44	33	22	11	0	11	22	33	44	55	66	77	88	600
700	51	38	26	13	0	13	26	38	51	64	77	90	103	700
800	59	44	29	15	0	15	29	44	59	73	88	103	117	800
900	66	49	33	16	0	16	33	49	66	82	99	115	132	900
1000	-73	-55	-37	-18	0	18	37	55	73	92	110	128	147	1000
1100	81	60	40	20	0	20	40	60	81	101	121	141	161	1100
1200	88	66	44	22	0	22	44	66	88	110	132	154	176	1200
1300	95	71	48	24	0	24	48	71	95	119	143	167	190	1300
1400	103	77	51	26	0	26	51	77	103	128	154	179	205	1400
1500	-110	-82	-55	-27	0	27	55	82	110	137	165	192	220	1500
1600	117	88	59	29	0	29	59	88	117	147	176	205	234	1600
1700	125	93	62	31	0	31	62	93	125	156	187	218	249	1700
1800	132	99	66	33	0	33	66	99	132	165	198	231	264	1800
1900	139	104	70	35	0	35	70	104	139	174	209	244	278	1900
2000	-147	-110	-73	-37	0	37	73	110	147	183	220	256	293	2000
2100	154	115	77	38	0	38	77	115	154	192	231	269	308	2100
2200	161	121	81	40	0	40	81	121	161	201	242	282	322	2200
2300	168	126	84	42	0	42	84	126	168	211	253	295	337	2300
2400	176	132	88	44	0	44	88	132	176	220	264	308	352	2400
2500	-183	-137	-92	-46	0	46	92	137	183	229	275	321	366	2500
2600	190	143	95	48	0	48	95	143	190	238	286	333	381	2600
2700	198	148	99	49	0	49	99	148	198	247	297	346	396	2700
2800	205	154	103	51	0	51	103	154	205	256	308	359	410	2800
2900	212	159	106	53	0	53	106	159	212	266	319	372	425	2900
3000	-220	-165	-110	-55	0	55	110	165	220	275	330	385	440	3000
3100	227	170	114	57	0	57	114	170	227	284	341	397	454	3100
3200	234	176	117	59	0	59	117	176	234	293	352	410	469	3200
3300	242	181	121	60	0	60	121	181	242	302	363	423	484	3300
3400	249	187	125	62	0	62	125	187	249	311	374	436	498	3400
	-20	-15	-10	-5	0	5	10	15	20	25	30	35	40	

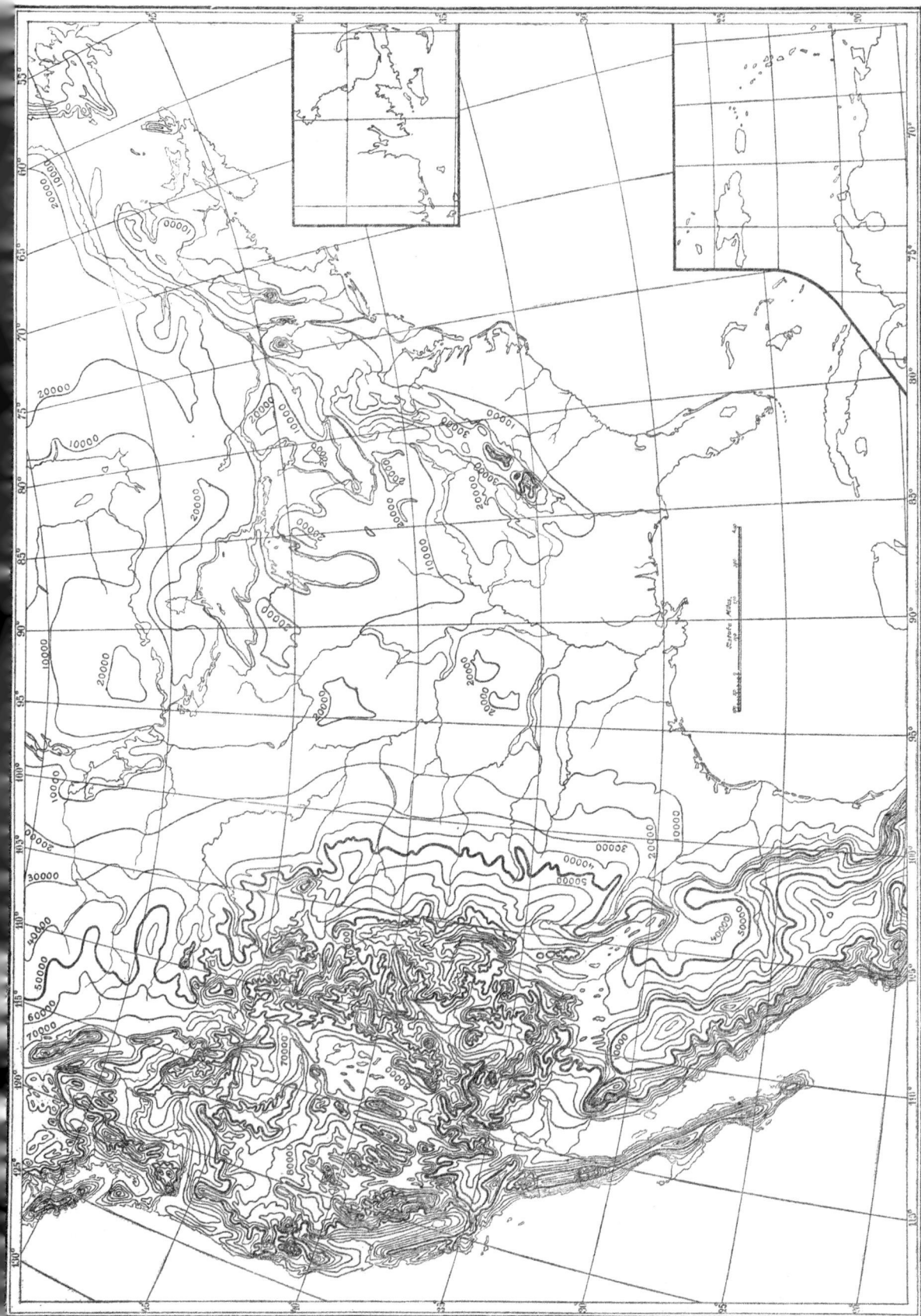

PLATE VIII. CHART OF LINES OF EQUAL POTENTIALS OF APPARENT GRAVITY AT THE SURFACE OF THE GROUND.